The Ingenious Mr Bell

The Ingenious

MR BELL

A LIFE OF HENRY BELL (1767 – 1830)
PIONEER OF STEAM NAVIGATION

BRIAN D OSBORNE

Argyll
publishing

First Published 1995
Argyll Publishing
Glendaruel
Argyll PA22 3AE
Scotland

Subsidised by the Scottish Arts Council

**British Library Cataloguing-in-Publication Data.
A catalogue record for this book is available from
the British Library.**

ISBN 1 874640 31 9

Origination
Cordfall Ltd, Glasgow

Printing
Images, Worcester

For my father
Malcolm M Osborne

CONTENTS

	LIST OF ILLUSTRATIONS	11
	ACKNOWLEDGEMENTS	13
	FOREWORD	15
1	TO SAIL BY THE POWER OF WIND, AIR & STEAM	19
2	"GAINING A THOROUGH KNOWLEDGE"	45
3	"WRIGHT IN GORBALS"	59
4	THE PROBLEM OF STEAM NAVIGATION	73
5	INN-KEEPER & PROVOST	95
6	"MR BELL'S SCHEME . . ."	113
7	STEAM BOAT PROPRIETOR	134
8	THE HEBRIDEAN & HIGHLAND STEAMSHIP	163
9	"HIS 'RULING PASSION' SCHEMING"	217
10	"BELL'S SCHEMING GOT BRITAIN STEAMING"	239
	Notes	265
	Sources	277
	Bibliography	279
	Index	281

LIST OF ILLUSTRATIONS

Page

14   *Helensburgh Quay* by unknown artist

22   *Comet* advertisement

25   *Comet*

29   *Comet* pasing Dumbarton Castle

       Plan of *Comet*

33   David Napier

       Bell's promissory note

49   Torphichen Mill

77   Jonathan Hull's steamboat

       Dalswinton Steamboat

85   William Symington

       Charlotte Dundas

93   Robert Fulton

94   Helensburgh Joseph Swan engaving

103   Henry Bell's business card

155   *Comet* on Forth Nasmyth

167   *Thames* in bad weather

189   Henry Bell letter

193   *Comet* account book

205   *Comet II* collision

ACKNOWLEDGEMENTS

As a librarian it gives me particular pleasure to be able to express my thanks for the courtesy and help extended to me by the many librarians and archivists who have assisted me in person or by answering written or telephone enquiries. In particular I would express my thanks to my ex-colleagues in Dumbarton District Libraries, especially Mike Taylor and Graham Hopner, for help in exploring the wealth of local resources held by them. I must also express my particular thanks to my colleague Don Martin of Strathkelvin District Libraries for many helpful comments and for leading me to some invaluable sources.

Many friends have provided me with advice, encouragement and photocopies and I would particularly wish to thank all who have borne so patiently with my enthusiasm for Bell and in particular would thank John and Olive Cubbage, Dave Harvie, Jessie MacLeod, Gerard Quail, Louis Stott and Mike Worling for their help, advice and support.

Brian D Osborne
March 1995

Helensburgh Quay (artist unkown). Thought to have been painted around 1830, shows a scene which would have been very familiar to Bell.
©*The Anderson Trust, Dumbarton District Libraries*

FOREWORD

One of the most striking facts about Henry Bell is that this book is the first full biography since Edward Morris's *The Life of Henry Bell* was published in 1844. Henry Bell made a major contribution to the early period of industrialisation. This son of a West Lothian millwright achieved what in James Watt's words, "many noblemen, gentlemen and engineers have puzzled their brains on" and failed to do. His status was acknowledged by contemporary engineers. "Bell did what we engineers failed in," wrote Isambard Kingdom Brunel. "He gave us the sea steamer; his scheming was Britain's steaming."

Yet to compare Bell's treatment by posterity with that of the American engineer Robert Fulton (1765-1815) is an instructive process. Fulton has, in the last century and this, been the subject of a wide and scholarly literature with British, American and Russian authors recording and interpreting his achievements. Bell and Fulton's achievements in steam navigation are certainly comparable and both men provoked strong reactions with enthusiastic supporters and bitter critics. However the Scots engineer has received only a fraction of the attention that has been devoted to his American contemporary.

One possible reason for this has been the lack of an accessible archive of Bell papers. Many of the documents which related to his steamship experiments and professional life would seem to have disappeared after his widow's death, although one remarkable survival did turn up in the course of the preparation of this book. In the United States a variety of libraries and archive offices hold a considerable quantity of Fulton material which has been available to be drawn on extensively by his various biographers and by those studying his technical achievements. Although there are in Scotland and England valuable sources, which have been used in writing this book, the documentary record relating to Henry Bell is much more sparse and information on Bell's life and activities has had to be gleaned, a detail at a time, from a very wide range of records.

Source problems cannot however solely explain the absence of any proper study of Henry Bell, nor can Edward Morris's *Life of Henry Bell* be considered to be of such excellence or so comprehensive a nature as to render further work redundant. Indeed Morris's book, though of considerable value, is quite uncritical and in any case devotes a large proportion of its space to an account of Morris's campaign on Bell's behalf.

The *Comet* centenary celebrations in 1912 might have been expected to be the occasion which would have seen the production of a definitive work on Bell. Although a substantial amount of coverage of the centenary and of Bell's life and work appeared in newspapers and periodicals, little of this was of much significance, and most clearly owed more to a re-cycling of Edward Morris than to any original research.

The lack of serious attention to Bell may owe something to the man himself. As will be seen, Bell was given to making inflated claims for his contribution to marine steam navigation, and it is possible that he has been his own worst enemy; that he has, by overstating his claims, brought about an under-valuation of his real worth. Bell, with his boasting, his shaky financial record, an artisanal background, his problems in literacy, was not perhaps the stuff of which Victorian hero figures were made.

It is also true that his life and work being, for the most part, centred on Glasgow and its environs, has been seen as provincial. While the origins of successful steam navigation in the British Isles are to be found on the River Clyde—and the later story of the Clyde as the world's pre-eminent shipbuilding centre needs no re-telling—there was a period in the nineteenth century, perhaps extending from the 1820s to the 1850s when the Thames was the British centre of shipbuilding and marine engineering. As the Clyde developed to rival the Thames there was a very great deal of rivalry between the two centres and the Thames shipbuilders and engineers lost no chance to denigrate the work of "country builders". This rivalry between English and Scottish engineers was not confined to shipbuilding. In such a climate of opinion Bell's role and contribution could perhaps not be expected to be favourably considered from an English viewpoint. Why Scottish writers have failed to give proper attention to a man whose achievement laid the foundation for an industry which won Scotland international renown is more obscure, and may need to be searched for among the darker recesses of the national psyche.

Having been brought up in Helensburgh, the town where Bell spent the latter half of his life and of which he was the first Provost, I was from an early age conscious of Bell and surrounded by relics and memorials. The flywheel of the *Comet* stood in the local park. The Baths Inn, built by Bell, still remained, albeit renamed the Queens Hotel. A monument to Bell dominated the sea-front. His head carved in stone looked down from the front door of the Municipal Buildings. The lack of any real memorial to Bell, in the shape of a serious biographical study, was however less evident.

In 1980, while preparing a local history publication on Helensburgh and the Gareloch area and conscious that that year was the 150th anniversary of Bell's death, I wished to include some information about Bell and his pioneering steamship. It quickly became clear that reliable information was hard to find and even such seemingly straightforward facts as the date of the launch of the *Comet* and her dimensions were hard to establish. What started out as a simple attempt to reconcile some of the conflicting information about Bell and the *Comet* has grown over the years into a search for a fuller understanding of a man who has been too easily written off as an accidental inventor, the "hero of a thousand blunders and one success", but who surely merits a considerably higher place in the story of marine engineering

17

and in the heritage of the nation. In the course of my researches into Bell he has grown from being an ostensibly well-known but, in fact, little understood local hero, to a fascinating, complex and significant figure. I hope that others will be equally intrigued by the story of the "ingenious Mr Henry Bell".

CHAPTER 1
To Sail by the Power of Wind, Air & Steam

The Glasgow newspapers of 1812 were not given to bold headlines or to sensational treatment of the news, but even for the densely packed four page broadsheets of the day the story of the arrival in the city of what was to be Europe's first successful sea-going steamship seems to be distinctly underplayed.

> We understand that a beautiful and commodious boat has just been finished, constructed to go by wind-power, and steam, for carrying passengers on the Clyde between Glasgow, Port Glasgow, Greenock and Gourock. On Thursday it arrived at the Broomielaw in three hours and a half from Port Glasgow.[1]

Thus the *Glasgow Herald* of Monday August 10th 1812 reported the coming to Glasgow of Henry Bell's pioneering paddle steamer *Comet* and with it, although the *Herald* writer could not be reasonably blamed for failing to foresee it, the dawning of the age of steam navigation. An age which would transform Glasgow and the Clyde and which would see the rise of a shipbuilding industry which would make the name of the Clyde famous throughout the world.

The few lines of restrained small print buried in the *Herald* doubtless reflect the fact that a steamboat, although a novelty, was not a new invention. The idea of steam power had attracted many experimenters and Bell was by no means the first to try to make a practical and commercial success of steam navigation. The well-informed Glasgow business men and country landowners who read the *Herald* would have heard of Patrick Miller's experiments in 1788 at Dalswinton Loch in Dumfriesshire and many might have seen the *Charlotte Dundas* coming into Port Dundas, Glasgow in 1803 when she was running her trials on the Forth and Clyde Canal. The Dalswinton steamboat and the *Charlotte Dundas* had demonstrated that steam propulsion was technically possible—the fact that no-one in Europe had taken matters any further forward in commercial application of the idea since 1803, when the *Charlotte Dundas* was laid up, might well have suggested to contemporary observers that a great, and perhaps unbridgeable, gulf lay between technical feasibility and practical application.

The most esteemed engineer of the age, James Watt, had given Henry Bell the benefit of his views on the matter. Bell, who had obtained an introduction to Watt from Glasgow's Lord Provost Gilbert Hamilton, had sought the great man's opinion on the application of the steam engine to ships. Watt, who through his partnership with Matthew Boulton and their control of the key patent for the separate condenser, had dominated the British steam engine field for twenty five years, replied to Bell and expressed the received wisdom of the professional engineering world in the discouraging words:

> How many noblemen, gentlemen, and engineers have puzzled their brains, and spent their thousands of pounds, and none of all these, nor yourself, have been able to bring the power of steam in navigation to a successful issue.[2]

These distinctly off-putting words from the doyen of British engineers should surely have persuaded Bell—who was neither a nobleman, a gentleman, nor in any sense that we would recognise the term today, a professionally trained engineer—to abandon his unrealistic dreams of steam navigation and stick to his trade as a house-builder. However they did not have the intended effect. To the chagrin of many of his contemporaries and, it would seem, of many commentators since, the ill-educated, semi-literate stone-mason and millwright succeeded where better-equipped, better-educated, better-funded and more skilled experimenters had failed.

Soon after the first *Glasgow Herald* story appeared, Bell's success was underlined by the first ever appearance in the British press of an advertisement for a steamboat service. Bell announced to the Glasgow public that the Steam Passage Boat *Comet* would leave Glasgow's Broomielaw Quay on Tuesdays, Thursdays and Saturdays at mid-day with return trips from Greenock departing on Monday, Wednesday and Friday mornings.

> The elegance, comfort, safety and speed of this Vessel requires only to be proved , to meet the approbation of the Public. . .

The name Bell chose for his pioneering vessel was a reference to a prominent astronomical comet seen in Scotland in the latter months of 1811. The popular notion of a comet as a fiery traveller through space made the name a more than apt one for a ship driven by fire.

One of Bell's main motives for pursuing the dream of steam navigation is revealed in a further paragraph in the advertisement:

> The Subscriber continues his Establishment at HELENS-BURGH BATHS, the same as for years past; and a Vessel will be in readiness to convey Passengers in the *Comet* from Greenock to Helensburgh.[3]

Bell, who in his forty five years had pursued a wide variety of occupations, was in 1812 landlord of the Baths Inn in Helensburgh, Dumbartonshire, on the north shore of the Clyde some twenty two miles down-river from Glasgow.

STEAM PASSAGE BOAT,
THE COMET,
Between Glasgow, Greenock, and Helensburgh,
FOR PASSENGERS ONLY.

THE Subscriber having, at much expence, fitted up a handsome VESSEL to ply upon the RIVER CLYDE BETWEEN GLASGOW and GREENOCK —to sail by the Power of Wind, Air, and Steam—he intends that the Vessel shall *leave the Broomielaw* on *Tuesdays, Thursdays and Saturdays,* about mid-day, or at such hour thereafter as may answer from the state of the Tide—and *to leave Greenock* on *Mondays, Wednesdays and Fridays,* in the morning, to suit the Tide.

The elegance, comfort, safety and speed of this Vessel requires only to be proved, to meet the approbation of the Public; and the Proprietor is determined to do every thing in his power to merit public encouragement.

The terms are, for the present, fixed at 4s. for the best Cabin, and 3s. the second—but, beyond these rates, nothing is to be allowed to servants, or any other person employed about the Vessel.

The Subscriber continues his Establishment at HELENSBURGH BATHS, the same as for years past; and a Vessel will be in readiness to convey Passengers in the Comet from Greenock to Helensburgh.

Passengers by the Comet will receive information of the hours of sailing, by applying at Mr. Houston's Office, Broomielaw; or Mr. Thomas Blackney's, East Quay Head, Greenock.

HENRY BELL.

Helensburgh Baths, 5th August, 1812.

The first ever advertisement in the British press for a steamboat service, placed by Henry Bell

22

The Baths Inn had been built by Bell a few years before to cater for the new fashion for sea-bathing which had started to have its impact on the area. The long and wearisome journey from Glasgow to Helensburgh, six hours by coach over rough and rutted roads, made the little village of Helensburgh somewhat remote for Glaswegians seeking a salt water cure or simply a restful holiday and restricted the development of Bell's hotel enterprise. Peter Mackenzie's *Old Reminiscences of Glasgow and the West of Scotland* tells of Bell's coach which served the Baths Inn and was driven by Thomas Bell, Henry's brother.

> That first coach of Henry Bell, guided by his brother Thomas, made the journey from Glasgow to Helensburgh, a distance by land of only some twenty-two miles, in the space of six hours: and good expeditious travelling it was thought to be in those days. . . This famous coach only started from Glasgow for Helensburgh during three days of the week, and then only in the summer months, for there was no coach of any kind on that road in the winter months. . . The very fatigue and tardiness of that journey, we can have no manner of doubt, impelled the genius of Henry Bell in regard to his steamboat. . .[4]

The alternative route was a journey down river by small passenger craft, the so-called fly-boats. This was equally unattractive being both lengthy and extremely uncertain. The ability of the steamship to make progress against the wind and currents brought both a shorter journey time and a degree of reliability to Clyde shipping which had until then been entirely absent. The development of the Clyde estuary as a residential and tourist location is almost entirely due to the coming of the steamship.

It is perhaps now hard for us to realise how little developed the River Clyde was in 1812. Although work had started in the 1770s on John Golborne's scheme to deepen the river by building groynes or projecting walls to increase the scouring effect of the water flow, this had still only had a limited effect. In the year of *Comet*'s maiden voyage Thomas Telford's 1806 plan to canalise the river by building a retaining wall joining up the ends of Golborne's groynes was almost complete. Despite such major efforts the trade of Glasgow was still almost all carried out through the down river outports of Port Glasgow and Greenock on the southern, Renfrewshire, shore of the Clyde.

The little *Comet* at 42 foot in length and drawing only 4 foot of water was about as large a vessel as could be assured of reaching the Broomielaw quay in the centre of the city. Even so her sailings had to be adjusted to take advantage of high water so that she might safely cross some of the shallowest stretches of a river which was notoriously obstructed by sandbanks and shoals. Indeed, we have from Henry Bell's wife, Margaret, an account of the stranding of the *Comet* on her first trip from Glasgow. Andrew McGeorge writing in *Old Glasgow; the place and the people* tells how:

> In 1812 our first steamer, the tiny *Comet*, with a draught of only four feet, grounded at Renfrew, although Henry Bell was careful to regulate her time of sailing so as to avoid low water. This was told to me by Mrs Bell, who said she was on board at the time. "And what was done then," I asked. "Oh," was the reply, "the men just stepped over the side and pushed her across the shoal.[5]

Another contemporary author recording his experiences on the river, "J. McN.", writing in James Pagan's *Glasgow Past and Present* wrote:

> There has been so much said about the Clyde, and the improvements which have been made upon it, that it is almost impossible to say anything new and true upon such a subject. I shall only state that I have sailed in the "fly-boat" from the Broomielaw to Greenock, and taken twelve hours to perform the feat; and also in the *Comet* of 1812, and

Comet *sails on the Clyde*
(Dumbarton District Libraries)

managed the task in six hours, although lying on the bank at
Erskine for a couple of them.[6]

If the Clyde was undeveloped, and no vessel large enough to
engage in foreign trade was to be able to come into the city centre
harbour at the Broomielaw until after 1818, the city itself was
still far from the condition it reached at its Victorian zenith as
the "Second City of the Empire". The Census of 1811 had recorded
a population of 110,460—a century later this figure had increased
almost ten-fold.

The contrast with Edinburgh is remarkable. In 1811
Edinburgh was slightly the larger of the two cities with a
population of 112,962 which grew fourfold in the century to a
1911 total of 423,464. Over the course of the next century Glasgow
was to outstrip the Scottish capital in size until it became two
and half times as populous.

Glasgow's rise to national and indeed international pre-
eminence was due to its development as a manufacturing centre.
This in turn was inextricably linked with the development of
the River Clyde, its dredging, widening and deepening. By the
height of the Victorian era the city boasted many miles of deep-
water docks accommodating ocean-going ships from all round
the world and the shipyards of the city produced a bewildering
range of ships of all sizes and types. The old cliché "The Clyde
made Glasgow and Glasgow made the Clyde" contains much
truth and not the least significant step in this process was the
coming of steam navigation to the river, a development signalled
by the advent of Henry Bell's *Comet*.

One might think that so momentous an event would be so
well documented and recorded that nothing more would now
require to be discovered or said about a ship which must rank
with the *Golden Hind*, *Victory* and *Titanic* in the catalogue of
historic British ships. However, as with so many other aspects of
Bell's life and work, the opposite is the case and much error and
contradictory information still exists. It must be admitted that
much of the blame for this confusion lies with Bell, who is not

always the most reliable source and who, particularly in his latter years, became increasingly prone to exaggerating his admittedly great contribution to steam navigation.

The first up-river voyage of the *Comet* from her builder's yard at Port Glasgow to the Broomielaw, reported in the *Glasgow Herald*, was on Thursday 6th August and her down-river maiden voyage took place in the following week. With the first advertisement appearing on Tuesday 11th August one may conclude that the maiden commercial voyage was on either Thursday 13th or Saturday 15th August.

Despite this many sources claim that the *Comet* was launched in 1811 or January 1812 and first sailed early in 1812 and Bell himself, at various times, added to the confusion by giving both 1811 and 1812 as the dates for the launch of his ship. Even the usually reliable contemporary historian and statistican James Cleland, a voluminous author who was the city's Superintendent of Public Works, writing in his *Rise and Progress of Glasgow* (1820) gives a January 1812 launch date. The plans of the *Comet* presented by her builder John Wood to the engineer David Napier are, to add confusion to confusion, endorsed "Built at Port Glasgow for Mr Henry Bell, 1811"! [7]

What such a launch date would fail to account for is the inordinate length of time between launch and the first recorded sailing. It is probable that the contract was signed in 1811 and the keel may even have been laid in that year.

There is however ample excellent, if seemingly little known evidence, for a July or August 1812 launch date. John Wood's yard supplied information on their steamboat production to a Government Committee in 1822 and stated in this that the *Comet* was launched on 24th July 1812.[8] To add to this the *Comet*'s engine builder, John Robertson, was interviewed many years later by the Glasgow antiquarian John Buchanan and gave him much specific and apparently accurate information on Bell and the *Comet*.[9] Robertson spoke of an August launch date with the engines on board and steam up. Another contemporary account from her first Master, Captain William MacKenzie, gives a slightly

27

different story with an account of a July launch. He recorded that:

> In March 1812 I was engaged by Mr Henry Bell of Baths, Helensburgh, to sail his *Comet* steamer as master, and on 1st April of same year was sent to superintend her building in Mr John Wood's ship building yard, Port Glasgow. Launched her in July with all her machinery on board, completely ready for sailing. Sailed early in August following with passengers between Glasgow and Greenock.[10]

Of these dates, the one supplied from the yard records, confirmed by the MacKenzie evidence, is perhaps more likely to be correct. The only possible reason for doubt is that if the ship was launched with steam up then there would be little cause for further delay before the maiden voyage and this might be thought to make a launch early in August more probable.

The long period from contract to launch for a forty two foot boat was in part due to Bell's endemic financial problems and in part perhaps due to the novel task of commissioning and assembling all the component parts—boiler, engines etc—from a variety of Clydeside engineers. As will be seen, in one case at least, two separate attempts at the boiler were required before a satisfactory product was delivered. The engine fitted to the *Comet* was a three horse power one which Robertson had on hand, in fact he had built it in 1811 "on spec". Robertson and Bell had been discussing steam navigation off and on since 1807 and both had seen Symington's *Charlotte Dundas* lying on the Forth and Clyde Canal. When, in 1811, Bell finally decided to go ahead with the *Comet* he bought Robertson's engine and had it shipped down river in May or June 1812 in a gabbart or lighter to be fitted at Wood's yard.

Some of the confusion which has, from her earliest days, surrounded the *Comet*, a confusion reflected in the very varied representations of her made by contemporary artists, is doubtless due to the changes which she underwent at various stages.

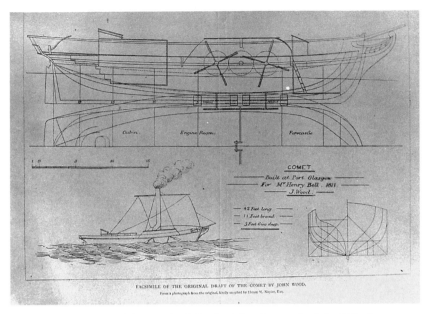

Facsimile of the original draft of the Comet *by John Wood (Mitchell Library, Wotherspoon Collection)*

The Comet *passing Dumbarton Castle. Note the different appearance of the funnel from the illustration on page 25.*

29

For example she was originally fitted with four paddle-wheels, arranged in pairs on either side. In practice these were found to be unsuccessful with the wash from each wheel adversely affecting its neighbour. Robertson had advised Bell of the likelihood of this before building but, as he said to John Buchanan over forty years later, Bell "was positive" and the ship was duly built with the double set of paddle wheels. They were soon replaced with what we now think of as the conventional pair of wheels.

The original engine was soon modified and was then replaced by a six horse power engine supplied by Thomas Hardie of Cartsdyke, Renfrewshire. Finally, in 1819, the ship was lengthened by James Nicol on the beach at Helensburgh. This somewhat odd expedient was undertaken, according to Robertson, partly to reduce the cost and partly because the obvious place to extend *Comet*, her builder's yard at Port Glasgow, was not readily available for the somewhat embarrassing reason that, seven years on, Bell still owed Wood part of the contract price. The lengthening, by some twenty feet, was undertaken to fit *Comet* better for Bell's ambitious plans to use her to run a West Highland service. These changes account for some of the many contradictory statements on size, engine power etc which are to be found in contemporary and more recent works.

Bell, although the designer, owner and in effect "inventor" of the *Comet*, utilised some of the leading talents in Clyde shipbuilding and engineering. The construction of her hull, having been entrusted to the noted Port Glasgow shipyard of John Wood, proved to be the foundation for that firm's early pre-eminence in steamboat construction. Between 1812 and October 1819 thirty nine steamboats were to go into service on the Clyde—and of this number no less than eighteen had come from John Wood's yard.

David Napier, of the Camlachie Foundry in Glasgow, had come into contact with Bell in the latter's role as a house-builder. Doubtless because of this connection he received the contract for the boiler for the *Comet*, a somewhat mixed privilege.

30

... I made the boiler and the castings for the engines of Mr Bell's little steamer the *Comet* for which he gave me his promissory note at 3 months which is still in my possession never having been paid. I recollect that we had considerable difficulties with the boiler, not having been accustomed to make boilers with internal flues, we made them first in cast iron but finding that would not do we tried our hand with malleable iron and ultimately succeeded with a liberal supply of horse dung in getting the boiler filled.[11]

The technique of using horse dung as a sealant was quite common—almost as common as issuing promissory notes and not paying up!

Napier carefully preserved among his papers Bell's promissory note for the boiler. This had obviously been renewed at least once. The extant version is dated 12th February 1814— about two years after the boiler was built—and promises to pay £62 to Napier three months after that date. It was renewed promptly—on 17th May 1814 and an extra eight shillings had been added to the account, presumably by way of interest. In July 1814 Bell managed to pay £35 to account and in August 1814 made two separate payments totalling £12.14.0. There is no record of any other payments to the balance of £14.14.0 ever having been made.[12]

Napier does not seem to have held this against Bell and indeed he was one of the Clyde engineers who lent their support to Edward Morris's campaign to get Bell official recognition and reward for his pioneering work.

Bell's shaky financial position and what must be seen as his somewhat doubtful business practices are also displayed in his dealings with the *Comet*'s engine builder John Robertson. Robertson was in business in Dempster Street, Glasgow, and in December 1812 Bell issued a promissory note, presumably for part of the work on the *Comet*'s engines, promising to pay Robertson £52 at three months notice. As was a common enough practice in a period before the full development of modern banking Robertson used the note to settle a debt of his own and endorsed it to a Kilmarnock merchant John Stevenson. He in turn endorsed it to a fellow-townsman, John Galbraith, who presented it for settlement to the Kilmarnock Bank. The Bank, finding Bell unable to settle the account, took a less tolerant line than David Napier and went through the ancient procedure, at that time the only practicable way in Scots Law of recovering a civil debt, of taking out a Decree of Horning against Bell and the other parties who had handled the note—a procedure which smacked more of the Middle Ages than of the dawn of the age of steam.

The Court of Session, in the King's name, charged Messengers at Arms and Sheriffs:

> . . . that on sight hereof ye pass and in our name and authority lawfully command and charge the said Henry Bell, John Robertson, John Stevenson, and John Galbraith to make payment conjointly and severally to our said Levite of the sum of Fifty Two pounds sterling and Interest thereof since due and till payment after the form and tenor of the said Bill registered Protest and Decree in all points within six days next after they are charged by you thereto under the pain of Rebellion and putting them to the Horn wherein if they fail the said span having elapsed that immediately thereafter ye denounce them as Rebels, put them to the Horn and use the whole other order against them prescribed by Law.[13]

The engine for the *Comet* proved to be insufficiently powerful, Bell having failed to take Robertson's advice.

To improve matters an enlarged cylinder was fitted when she was taken out of service to have her paddle wheels changed. The contract price for the original engine was £165 and the total cost, including the new cylinder and other modifications, was £365. It is reported that Robertson was never fully paid for either bill.[14]

What sort of vessel was this, the first of the long and distinguished line of Clyde paddle steamers? When re-registered after lengthening she was officially described as:

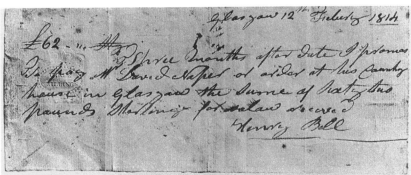

(top) David Napier who made the boiler for the Comet
(bottom) The promissory note made by Bell to Napier for the payment of the boiler. (Dumbarton District Libraries)

carvel built ... with a flush deck being furnished with a steam
engine and its machinery with which she sails.[15]

However the new age of steam had not yet done away with sail
and *Comet*'s iron funnel also served as a mast on which a square
sail could be hoisted to take advantage of any favourable winds.
Robertson reported that, whatever *Comet*'s initial mechanical
problems with double paddle wheels and underpowered engine
might be, she was, thanks to John Wood's skill as a designer, a
good performer under sail.[16]

Despite Bell's problems in paying Wood and sub-contractors
like Napier and Robertson he evidently did not skimp on the
finish or equipment of his steamer. John Robertson described
her as being

> ... prettily painted of different colours. She had a small figure
> head of a lady with red cheeks and coloured dress.[17]

In 1813 Henry Bell published a pamphlet *Observations on the
Utility of Applying Steam Engines to Vessels etc* and in this he gives
the best account of the internal arrangements of the *Comet*.

> ... in the year 1812, I built a vessel 40 feet long, and $10\frac{1}{2}$ feet
> beam and hold eight feet deep, which I fitted up solely for
> Passengers in the following manner:
> 1st In the stern there is a cock-pit of 6 feet, seated all
> round, with six lockers, neatly fitted up. On each side there
> is a stair, by which you ascend to the cabin.
> 2d The cabin is 10 feet 4 in length, and 7 feet 6 in breadth,
> elegantly furnished with sofas all round, &c. There are, also,
> moveable divisions, by which it can be divided at pleasure
> into three different apartments, two of which contain each
> two handsome beds. The third is formed into a small neat
> cabin, with seven lockers for holding stores, &c.
> 3d Next to the cabin is an apartment of 12 feet 6 in length,
> and 10 feet 6 in breadth, in which the engine and machinery
> are placed. And in the steerage, (which is 10 feet in length,
> and 7 feet in breadth) are 4 beds, two on each side, has also

six lockers, and seated all round.

The upper deck above cabin and steerage is seated all round, on each side there is a projection of 1 foot ten inches, being a recess for the paddles, which form part of the deck; and in each projection are a water closet.

This vessel has been running between Glasgow and Greenock for 6 months past, and is found very commodious for passengers.[18]

John Robertson's reminiscences as noted down by John Buchanan confirm some of these homely details of the little ship's internal arrangements:

> The cabin was entered at the back (next the helmsman) with two little flights of steps down on each side. The cabin which was very small had seats fitted in all round.[19]

The *Comet* was, according to John Robertson, crewed by four men. He lists the members of the first crew as being William MacKenzie, the Captain (he describes him as having run a small school at Helensburgh and knowing little of vessels); an engineer, Robert Robertson (in fact John Robertson's brother); the pilot, a Highlander called Duncan McInnes and a fireman—whose name is not recorded. No sailors were carried as none were needed.

The Highland pilot was the first of what was to be a long line of men from the West Highlands who made a career on the Clyde steamers. By the time the lengthened *Comet* was trading to Fort William in 1819 and 1820 we are told that the crew had increased to six men and a piper.

As can be seen this was a handsomely and really remarkably well-equipped vessel. She was indeed perhaps over-equipped for the four and a half hour voyage from Glasgow to Greenock, a passage which would hardly warrant the use of the "two handsome beds". It seems clear that Bell's imagination had already gone well beyond a simple ferry service from Glasgow to Greenock with connection to Helensburgh for the Baths Inn.

35

If we may judge by the facilities provided when Bell drew up the plans for the *Comet*, he already had in mind longer passages which would require overnight accommodation.

We have seen that the original engine was quickly modified to produce more power and Bell himself admitted in his pamphlet that the *Comet* was underpowered:

> The engine is a small portable one, of only 3 horse power; a vessel of her size would require an engine of 5 horse power, by which she could run the same distance in three and a half hours.[20]

Important as the elegant cabin lamps and handsome beds might be in enticing passengers on board and winning public confidence in this strange new means of transport the *Comet*'s success would, in the end, depend on the degree of reliability and safety she could offer and the speedy passages she could afford. Bell made bold and substantial claims for his brainchild.

> In this country many attempts have been made to apply steam engines to the purpose of driving vessels; and vessels have been set agoing, and even patents obtained for them, but never one of them was found to answer the end.
>
> The above vessel is the first that ever answered the purpose; the engine of which is so constructed, that no change of position can have any effect upon it. It goes as well when the vessel is tossed with a heavy gale, as in a calm. This has been experienced; for the *Comet* . . . has run between the island of Bute and Greenock in very stormy weather, with high seas, which is known to be a more dangerous passage than out in the open ocean.[21]

The success of the *Comet* was testified to by the speed with which imitators quickly appeared. The *Elizabeth*, engined by John Thomson followed in December 1812, the *Clyde*, with engines by John Robertson was launched in February 1813 and Bell himself produced the first, though unsuccessful, engine for the river's fourth steamer *Glasgow*—all incidentally the products of John Wood's yard.[22] The new technology created a new trade and a new level of interest in travel. Cleland, writing in his *Annals of Glasgow* in 1816, states:

> It has been calculated, that, previous to the erection of Steam-Boats, not more than fifty persons passed and repassed from Glasgow to Greenock in one day; whereas, it is now supposed that there are from four to five hundred passes and repasses in the same period.

Writing of these new steamboats he goes on to note that:

> The cabin and steerage are fitted up with every suitable convenience; the former is provided with interesting books, and the various periodical publications. . .
>
> Since the *Comet* began to ply on the River, it is very common to make the voyage of Campbeltown, Inveraray, or the Kyles of Bute, and return to Glasgow on the following day. . .[23]

Henry Bell had just invented what was to become the great Glasgow tradition of a trip "doon the watter." Peter Mackenzie's *Reminiscences of Glasgow* tells of the undeveloped age before the coming of the *Comet* and the start of the mass tourist industry which she engendered:

> . . . we heard old Provost Graham of Whitehill describe the fact, that when he first went from Glasgow to 'the saut water', it was to Helensburgh, when there were only 'two or three bits of houses in the place'. . . They could get neither sugar nor tea in the village; no butcher and no baker, no church, no surgeon or apothecary.[24]

Every new industry spawns its subsidiary or service industries and the steamboat was to be no exception. Where there is a tourist

37

there surely has to be a tourist guide-book. However in the nineteenth century it was possible to produce a guide-book in verse, and in 1824 there was published *The Steam-Boat Traveller's Remembrancer; containing, Poems descriptive of the Principal Watering-Places Visited by the Steam-Boats from Glasgow*. This work was written by William Harriston, described as a "weaver, soldier, fisherman and poet" and among its many truly dreadful verses are the following in praise:

Of Helensburgh and the Country Adjacent

There lives Henry Bell, so much fam'd
For exerting mechanical skill,
Who the first of our Steam-vessels fram'd-
Thus he merits the nation's good-will.

Far advanc'd in old age is the sire
Of the Steam-boats in Scotland, yet he
Retains a great share of the fire
Of activity, humour and glee.

Though infirm, yet he holds on the paths
Of Business; and here I may tell,
His House has some elegant Baths,
A commodious Inn and Hotel.

Disappointment's discouraging pangs
He often has suffer'd in part,
But Despair with its terrible fangs
Never yet found its way to his heart.[25]

One hesitates to commend the judgement of an author so obviously ungifted, but for all its deficiencies Harriston's doggerel is of some significance by giving us an insight into one, and possibly a typical, contemporary's view of Bell and his place in history. A less kindly commentator described Bell as the "hero of a thousand blunders and one success". Bell's friend and biographer Edward Morris weighed in on the side of Harriston and was, one may feel somewhat unfortunately, moved to poetry by the death of his hero Bell. Morris puts Henry Bell's claims to regard and recognition in the very highest terms.

The fire of genius came from heaven
From Him who David's harp inspir'd—
In love to man, in kindness given,
To raise our world,—Bell's mind was fir'd
With ardent light, mankind to bless;
And patriots will this truth confess.

The *Comet* moves,—Dumbarton's rock
Displays its front amid the storm—
She rides, nor heeds the tempest's shock,
A fairy thing, a beauteous form;
She triumphs on that trying day,
While shouts of joy burst on her way.

Port-Glasgow, Greenock, now behold,
The bark which battl'd wind and wave;
And Helensburgh, whose flags so bold
Were rais'd on high, to Bell the brave;
And his fam'd *Comet*, which had won
Renown afar, from sun to sun.[26]

The remaining seventeen verses of this epic are in a similar heroic vein.

Bell's life and the novelty of the steamboat would seem to have had a remarkable attraction for poets—these lines by "Britannicus" apparently appeared in the *Greenock Advertiser* around 1830 and are quoted by Morris in his *Life of Henry Bell*.

To the Genius of Helensburgh

Who conquer'd thee, Clyde! 'mid the winter's wild roar
And o'er thy dark bosom, 'gainst tides did explore
Lone spots, which thy billows do lave?
Who brav'd thy rude torrents, rode hurricane's blast,
And ocean subdu'd, when with tempests o'ercast?
It was Bell with his bark on thy wave.
The *Comet* he plann'd;—yes, the scheme was his own!
It will spread through the world, it will send his renown
O'er rivers and oceans afar.
He fought the hard battle, where many were slain,
And triumph'd o'er Neptune, through all his domain;
What millions his bounties now share![27]

Bell, as we can see, attracted widely different views. Some contemporaries like Harriston and Morris and "Britannicus" saw him as a figure of genuinely heroic status, while others characterised him as an ill-educated, vainglorious boaster and blunderer. Some saw him as a major engineering figure, while others then and since have pointed out the many defects in his work and his lack of any advanced technical training or grasp of the theoretical foundations of the fields he worked in. A number of these critics had a vested interest in denigrating Bell and in seeking to ensure that any official reward for the invention of the steamboat went elsewhere. It is thus interesting and instructive to get a professional view on this and to read the comments of a distinguished Glasgow engineer and contemporary of Bell's, James Cook, who, writing to James Cleland in 1825 said:

> I beg leave to state, that there is very little difference in the principle or construction of that kind of machinery in general use at present, from that applied by Mr Henry Bell, in his Steam Boat, *Comet*, erected by him in 1811 or 1812. The greatest improvement that has taken place since that period,

is in the construction of the boilers. . . It no doubt will be said, that other great improvements must have taken place, since the days of the *Comet*, as the speed of the vessels is greatly increased now to what it was at that time. This is allowed, but it does not follow as a matter of course, that this is due to some great improvement in the principle and construction, this has arisen from practical observations; those concerned with such undertakings being now better able to proportion the power to the size of the body to be impelled. The best possible proof that I can adduce in support of this observation, is the *Glasgow* Steam Boat, which boat by the bye, was built by Mr Bell's directions, in 1812, or 1813. The Engine and impelling machinery, were made and put into the vessel by me, in 1813, or 1814. The vessel, I believe, was lengthened a little since, to give accommodation, the Engine and machinery are still the same, and there are not many boats on the river at this day, that exceed her far in point of speed in still water. I do not recollect now what kind of speed she went at, but if it was slow, I am inclined to think the cause of that, was the want of a proper proportion betwixt the size of the vessel, and the power of the Engine and impelling machinery, and not owing to any defect in the principle or construction of the machinery. . .[28]

Quite apart from the merit of his work and the question of his technical ability, Bell was to be accused of stealing other men's ideas. He in turn unblushingly claimed for himself the credit for putting Robert Fulton, the American steamboat pioneer, on the right lines. A controversial figure in life, even in death Henry Bell provoked surprisingly sharp comment. When he died in 1830 the obituary writer of the *Glasgow Chronicle* noted with regret the death of:

. . . the ingenious Mr Henry Bell, the practical introducer of steam navigation into Europe. His constitution had been greatly broken down for many years, and his bodily sufferings were frequently very great. His "ruling passion",

> scheming, was strong to the last, and the advice of his best
> friends failed to check this impetuous disposition. It involved
> him in many difficulties, but the benefits derived from his
> successful experiments in the steam propelling system are
> of such magnitude as to ensure a grateful remembrance of
> his name. . .[29]

This is a remarkably frank assessment indeed for obituary writing, where one more usually finds that Dr Johnson's observation on memorials is generally adhered to: "In lapidary inscriptions a man is not on oath".

The *Chronicle*'s assessment of Bell's character and contribution is, despite its somewhat acid tone, perhaps a fairer summary than either the hagiographic view espoused by Morris or the "hero of a thousand blunders and one success" approach.

Although John Robertson, the builder of *Comet*'s engine described Bell as:

> . . . restless, unmethodical and by no means a good hand at machinery[30]

there can be no doubt that Bell was an ambitious and talented, if admittedly erratic, man who was much more than just the owner of the *Comet*. Perhaps due to his own over-stated claims— and what must be admitted was, at times, unattractive boasting Bell has never received quite as much credit as he is surely due.

Remarkably enough for such a well-known and significant figure his full story has never been previously told. The only approach to an account of his life, Edward Morris's *The Life of Henry Bell*, was published in 1844. This work had the considerable benefit of being written by a personal friend of Bell. Over the last four years of Bell's life Morris received much detailed information from Henry Bell and his wife Margaret and had access to, now vanished, private papers. Despite these advantages Morris's book is remarkably unsatisfactory as a biography, concentrating almost as much on Edward Morris's attempts to win for Bell a Government pension or grant as it does on the subject's life. It also suffers from Morris's quite uncritical acceptance of everything that Bell told him.

How the son of a West Lothian millwright came to attempt what, in James Watt's words, "many noblemen, gentlemen, and engineers have puzzled their brains, and spent their thousands of pounds" on, and how he first, despite many setbacks and disadvantages, managed to be "able to bring the power of steam in navigation to a successful issue" is a more complex story than might be surmised from many of the somewhat dismissive references to Henry Bell in the 165 years since his death. The story takes in such elements as his very widely based practical training in masons and millwright work, shipbuilding and general engineering, including a spell working for the great Scots engineer John Rennie in London, the establishment of a sucessful builder's business in Glasgow, a three year term as Provost of Helensburgh, a mass of plans for canals, water supply schemes and reservoirs, his part in the opening up of the Highlands and Islands of Scotland to steam navigation, trade and the tourist and even, perhaps inevitably, a secret weapon!

Although some of Bell's rivals among his contemporaries were highly critical of him and took every opportunity to disparage him he nonetheless must rank as one of the major figures of the period. His true status was indeed freely acknowledged by some of the leading engineers of his day. Isambard Kingdom Brunel, chief engineer of the Great Western Railway, builder of the *Great Eastern* and the *Great Britain*, an engineer of international reputation, was one who recognised Bell's part in the creation of the new age of steam. Brunel contributed generously to the fund established to provide support for the sick and penurious Bell towards the end of his life with the apt words:

> Bell did what we engineers all failed in—he gave us the sea steamer; his scheming was Britain's steaming.

43

and said to Edward Morris when he sought his support for the fund:

> . . . the British government should have given Mr Bell £1000 per annum, twenty years ago. . . He worked nobly for his country; he accomplished what others had failed in.[31]

CHAPTER 2
"Gaining a Thorough Knowledge"

The fast-flowing River Avon forms part of the boundary between West Lothian and Stirlingshire. Along the Avon's banks in the West Lothian (or as it was known in the eighteenth century Linlithgowshire) parishes of Torphichen and Linlithgow and the Stirlingshire parishes of Muiravonside and Polmont, were once to be found a great many water powered mills. The parish minister of Muiravonside, writing in the *First Statistical Account* in the 1790s recorded no less than 17 such mills on the nine and a quarter mile stretch of the Avon contained within his parish. These establishments included flour, corn, barley, flax and lint mills and the operation, building and maintenance of such mills was a highly skilled trade which had, for several generations, profitably occupied the Bell family. Henry Bell was to write:

> The Bells, as millwrights, were known, not only in Scotland, but in England and Ireland.[1]

However, in addition to millwrighting the Bell family had been engaged as builders, wrights and engineers on projects around their East Central Scotland base.

It is not clear where the Bell family originally came from; although Henry Bell, writing to his biographer Edward Morris in October 1826, confidently stated:

45

> I may refer you to McLure's History of Glasgow, where great
> improvements performed by my ancestors, the Bells, in that
> city are fully recorded.[2]

The work to which he directs Morris is in fact McUre's, not
McLure's, *History of Glasgow*. But there is no indication in Bell's
letter of which part of McUre's work he had in mind. The section
quoted below, from the 1830 edition, is the obvious reference
and is indeed the only part of the work which could be construed
as having a connection with the Bell family.

> The names of the four eminent provosts, that contributed
> most to the advantage and beauty of the city. The first was
> Sir Patrick Bell who in his time caus'd build the Townhouse
> and Tolbooth, with the stately steeple thereof, the Meal-
> market, Fleshmarket, and the Tronesteeple, all built the time
> he was provost.
>
> Sir John Bell, his son, was the second provost, a worthy
> man, he caus'd build the Gild-hall and the stately steeple
> thereof, and in November 1677, when the great fire in the
> city brake out that upwards of a thousand families, and one
> hundred and thirty shops and houses were consumed to
> ashes, so immediately thereafter he encouraged the
> proprietars of the waste houses to rebuild their houses in a
> more magnificent structure than ever it was before, and gave
> them vast sums of money furth of the revenues of the city
> for advancing the work which is admired by all strangers.[3]

Patrick Bell, who despite McUre's assertion, was not a knight,
was Provost in the 1630s and his son, Sir John Bell had no less
than ten periods of office between 1658 and 1681.

Whether there was more than just a handed down family
tradition behind Henry Bell's belief that he was descended from
these notable Lords Provost of Glasgow is difficult to say. It
should however be noted that the Christian names Patrick and
John were both still current in the Bell family in Henry Bell's
generation. Scottish traditional naming customs were still strong

and family Christian names can often be traced back for many generations. While John is certainly a common enough name in the period, Patrick is perhaps just sufficiently less common to lend some credence to the belief that the West Lothian miller's family were descendants of these distinguished Glasgow provosts.

Henry Bell's father, Patrick Bell, was born in 1727 and in December 1750 married, in Linlithgow Parish, Margaret, the daughter of John Easton of Wester Inn, Stirlingshire. Patrick and Margaret had the typically large family of the period, Margaret giving birth to six sons and four daughters. Of these ten children only four—Henry, Thomas (who we have already encountered as the coachman of Henry's Glasgow to Helensburgh coach), Margaret and Elizabeth—outlived their father, who died in 1793 aged 66.

Margaret Easton came from a family with a similar background to Patrick Bell's and the Eastons, according to Henry, were also well known builders. Notable members of the family included Alexander Easton, employed as a resident engineer, under Thomas Telford, on the Caledonian Canal. Alexander had also earlier worked as a mason on the Forth and Clyde Canal.

To judge from the surviving records in the Old Parish Registers Patrick and Margaret obviously moved around to some extent in the earlier years of their marriage. Their sons Patrick and John were recorded as being christened in Bathgate parish in 1753 and 1756, at a time when their father was the lessee of the Birkinshaw mill near Bathgate. Patrick Bell gave up this property in 1765 and moved his family some twelve miles, to the Torphichen Mill, about half a mile from the village of Torphichen in West Lothian.

The village lay at the centre of a prosperous agricultural parish whose chief claim to fame lay in the remains, next to the parish church, of the preceptory of the Order of the Knights of St John of Jerusalem. Torphichen had been, from the 12th century to the Reformation, the Scottish headquarters of this crusading order. Towards the end of Patrick Bell's life, in 1789, he moved from Torphichen to take over the mills of Brigh.

While at Torphichen Patrick and Margaret's fifth son, Henry, was born on 7th April 1767. Young Henry was in due course sent to the village school. In an unpublished and now apparently lost memoir, which is referred to by his biographer Edward Morris, Bell says that after being at school for two years he decided to leave and hired himself out to herd cattle. But with the onset of winter he was glad enough to return to the village school and settle again to his studies. Life at Torphichen was presumably congenial to Henry because we are told by the Rev William Hetherington in the *New Statistical Account* of the parish that:

> A few years before his death he paid a visit to the secluded scene of his infancy, with a view, it is said, of purchasing the spot, and erecting another cottage, and in that calm retreat terminating a career, the honours and rewards of which had been, as too often happpen, but ill-proportioned to its usefulness. He did not, however, prosecute his intention, and the solitary ruin remains his melancholy memorial.[4]

The aspirations of the Bell family for this younger son and the skills and training they thought were necessary if young Henry was to follow in the family trade of millwright are perhaps hinted at by the decision taken by his parents in April 1776. They took him away from the local parish school in Torphichen, at the tender age of nine, and sent him to lodge with his mother's uncle and aunt in Falkirk so that he could attend school there. It would seem that the attraction of the burgh school in Falkirk to his parents was the superior quality of mathematical teaching offered there by the school-master, one Mr Shaw. The young Henry was to spend over three years at this school and the technical education he received there was obviously to stand him in good stead in his later career.

Unfortunately Bell's education served him less well in the area of writing, grammar and spelling. Although in his later life

48

The ruin of Torphichen Mill, Bell's birthplace in West Lothian
(West Lothian District Libraries)

he was to publish various pamphlets and broadsheets and have numerous letters published in the press to promote his many schemes, these all bear the mark of another hand, perhaps his wife's. If they had been received in their original form from Bell's hand they would certainly have required considerable editorial effort to make them intelligible. In 1827, while sparring over the primacy of the development of steam navigation with Patrick Miller Jnr, Miller had made mockery of Bell's spelling and grammar and his comments provoked the following dignified response from Bell:

> It is a pitiful display of weakness and malice in Mr Miller, to chuckle over a letter of mine, in which he finds bad spelling, and incorrect grammar, thinking by this to produce contempt, and want of confidence in my rightful claims. . . I could quote many instances of engineers, and others, who, like myself, had a scanty education previously to their advancement in the world, which they felt, and lamented in after life; but still this deficiency prevented neither them nor myself from accomplishing for our country's good, that which many a good penman, and finished grammarian, could never have achieved. . .[5]

Bell's comment is just, and although these dignified words doubtless expressed his genuine and heartfelt sentiments they were certainly not all his own work. Edward Morris noted that:

> . . . to the day of his death he felt the great want of being unable to convey his ideas in language which cultivated and polite society approve of. There was a richness and genius in his thoughts and communications, but the rough garb in which these were arrayed was to his disadvantage, especially with those persons (and there are too many of them in the world) who look more on the surface than into the depth of things.[6]

Quite how far into "the depth of things" one had at times to look to decipher Bell's meaning in his writings is indicated by the following verbatim extract from a manuscript letter written by Bell in 1820 to Dr Archibald Wright regarding the sale of shares in the *Comet*:

> I was favered with a letter from Mr McIntayer—informing me that you wished to have a shair in the *Comet* steam boat— which you shall have as follows one shair is just now £55 pounds but you deduct 20 persent from a fifty pound shair as original is £10 which leaves to be paid 45 pounds and you will have a right to an equal dividend of what she makes from the first of August 1820—this is my terms and the said £45 to be paid me just now and for surety of your I cass you to be indorsed on the back of the register and allow on the general rindishion of the *Comet* steam boat and your rights will be as good as the other partners if this is agreeable to your views pleas send me the cash.[7]

Even in publications and letters where the assistance of a more "finished grammarian" can quite easily be detected the typically breathless style of Bell shows through. In his manuscript letters the flow of ideas, energy and imagination is impressive and the character of the author is obvious. However the reader's understanding is impeded by poor handwriting, eccentric spelling, almost non-existent punctuation and shaky understanding of grammar and syntax. Nonetheless Bell's other correspondents, like the Dr Wright addressed in the letter quoted above, clearly managed to understand what Bell meant and his draft for £45 in payment of a share in the *Comet* was received within the week.

His formal education at Falkirk was completed at the age of thirteen. In 1780 the young Henry made his start in one of the established family trades and commenced work as a stone mason with one of his relatives. He was to spend three years working at this trade before changing to what was perhaps a more technically demanding occupation, one more central to the family tradition, and one which would make good use of the technical and mathematical skills acquired at Falkirk burgh school. Around 1783 he became apprenticed to his uncle Henry Bell, of the Jay Mill, to learn the trade of millwright.

Millwrights, many of whom combined the operation of a mill of their own with the construction and maintenance of mills for others, were among the most highly skilled of the rural tradesmen. Their craft involved a mastery of the varied skills of building, metalwork and woodwork coupled with the problems that required to be overcome in the harnessing of waterpower. The larger undertakings such as urban flour mills were often major construction projects, indeed they were among the largest such projects of the period, and their design and execution represented a significant architectural, engineering and administrative challenge.

With the formidable range of technical and managerial skills demanded by the trade it is not surprising that many of the leading engineers of Bell's era were to spring from the ranks of the millwrights and millers. In particular the demands of the new industries, techniques and processes of the Industrial Revolution seemed often to be best met by the multi-disciplinary talents of the millwright.

Among the most notable of these millwrights turned engineering consultants was John Rennie (1761-1821). Rennie began a distinguished career, which ended with his becoming one of the leading civil engineers of the early nineteenth century, by being apprenticed in 1773 to Andrew Meikle at the Houston Mill in East Lothian. Andrew Meikle was to win fame as the inventor of the threshing mill, patented a corn-dressing machine and introduced many significant improvements to the working of his mills. He came from a notable dynasty of millers and millwrights renowned throughout lowland Scotland for a century and more. His apprentice's background was somewhat different. John Rennie's family were small lairds, indeed Meikle's mill stood on the Rennie estate of Phantassie, and an apprenticeship to a mechanical craft was certainly not the usual or expected career for even the younger son of an eighteenth century landowner to follow. However the millwright's trade was one which was obviously both congenial to the young Rennie's technical bent and, as importantly, one which was seen as being of sufficient status to satisfy his parents' ambitions.

By the age of nineteen Henry Bell had gained a practical knowledge of stone masons' work and had served his time as an apprentice millwright. With this varied training and a network of Bell and Easton relatives well established in the prosperous and increasingly industrialised Lothian and Stirlingshire area, an obvious career progression would have been open to him. He could be considered to have been fortunate enough to have had, by both family background and training, an excellent opportunity to enter, without obvious difficulty, into a most respectable and thriving trade. However the young man would seem to have had, from an early period, the idea that his life was to not to follow in the settled paths of the Bell family tradition and in 1786 he turned his back on a life as a miller or millwright and went off to nearby Bo'ness to start work in the shipyards there.

Bo'ness (Borrowstounness to give it its full form) was in the 1780s a significant port. Lying on the Firth of Forth in West Lothian it was in fact Scotland's third most important seaport with an extensive trade being carried on with Scandinavia, the Low Countries and France. The export of coal and salt was the staple trade of the port and was combined with the importation of timber and other marine stores from Scandinavia and the Baltic. The disruption to the West Coast trade in timber and timber products with the Americas caused by the American War of Independence between 1775 and 1783 had increased the importance of the East coast harbours like Bo'ness as ports of entry for timber, pitch, hemp, flax and other essential naval stores. Bo'ness was however to suffer something of a decline due to the East coast terminus of the Forth and Clyde canal being established a few miles away at Grangemouth and the consequent channeling of much trade through the rival port.

The Bell family had long had connections with Bo'ness, one

branch of the family having been tenants of Kinneil Mill in the vicinity of the town. Apart from its shipping interests Bo'ness lay at the heart of one of the most important centres of the Industrial Revolution in Scotland and the area saw some of the earliest applications of steam engines. Kinneil Ironworks on the outskirts of Bo'ness, under the management of Dr John Roebuck of the Carron Company, was an early location for a Watt engine and James Watt had carried out various experiments in steam power in a workshop in Kinneil Woods. An important pottery had been established in 1784 and soap and chemical works were also operating in the town, along with a silk spinning industry. Henry Bell had thus moved to live and work in an area which, for a technically minded young man, must have offered a great deal of stimulus and excitement.

The shipbuilding industry had become established at Bo'ness in the mid eighteenth century and by the time Bell moved there two firms of shipbuilders were operating. These were the yard of Thomas Boag and the yard where Bell was to find work, Shaw and Hart. Both were ocuppied with building ships of between 40 and 350 tons for service in the coasting and continental trade.

Bell himself told Edward Morris that his object in going to Bo'ness, to enter the employment of Messrs Shaw and Hart was:

> ... for the purpose of being instructed in ship-modelling. I wrought with them in ship-work for the space of one year.[8]

This year at Bo'ness would therefore seem to have not been simply an attempt to establish for himself a new career in shipbuilding but rather a more calculated and deliberate attempt to widen his horizons and acquire a new range of technical skills. Henry Bell at the age of nineteen or twenty was, it would seem, a clear-sighted and ambitious young man who was setting out to make a career for himself and was determined to put together for himself the elements of a wide-ranging and sound practical training.

There was in Bell's day no recognised academic training for engineering. A craft apprenticeship followed by a period of work as a journeyman was the regular route adopted by those intending to set up in business as independent master craftsmen or, more ambitiously, as civil engineers like John Rennie. Most such aspirants would have confined their training to one branch of engineering. A very few, like Rennie, when established in business, undertook such university courses as were available and which seemed appropriate to their interests. However this was a very exceptional procedure and most recruits to engineering never came within the ambit of the university system. This was particularly true in England where the two universities of Oxford and Cambridge were only open to members of the Church of England, and were very restricted in their syllabus. The Scottish universities as a result did tend to attract both Dissenters from south of the Border and those wishing an education less narrowly based than that available in the Oxbridge colleges. Even so, few of Rennie's professional contemporaries followed his route. Rennie's attendance at University is of course less surprising in view of his middle class background.

Ship-modelling, or the design of ships, was quite far removed from the earlier lines of work Bell had followed and was certainly not the most obviously useful skill that an aspiring young engineer with a background as a millwright might see the need for. Whether Bell, as early as 1786, had in his mind some thoughts of making a career in shipbuilding or shipowning is not known but the knowledge he gained during his time with Shaw and Hart certainly was to prove of use to him later. He would seem to have established a good relationship with his employers because, many years later, it was to Bo'ness, to Shaw and Hart's yard, that he brought the *Comet*, making what was incidentally the first operational passage of a steamboat through the Forth and Clyde Canal, for a refit in 1813.

His year at Bo'ness over, Bell next went to the Lanarkshire coalfield, where he spent a year in the employ of an engineer,

James Inglis, at Bellshill, near Motherwell. There is very little known of this period in Bell's life, he himself explains his motives for the move as being:

> . . . for the purpose of gaining a thorough knowledge in the engineering.[9]

The coal-mining and iron industry of the area would offer Bell yet another valuable range of experience and training and these moves all seem to add up to a careful and considered plan on Bell's part to equip himself for an ambitious career. If an apprenticeship to a millwright, a year working on ship-modelling and a year in an engineer's shop testified to Bell's grand design the next step was undoubtedly exceptional. Bell moved to London and for eighteen months from 1788 worked for John Rennie.

Rennie at this time was in his late twenties. Although not yet the dominant figure he was to become, he was already a force to be reckoned with in engineering circles, not least because of his practical and theoretical training and ability. Bell spoke of this period with justifiable, pride:

> After fulfilling my time with Mr Inglis, I went to London, to the famous Mr Rennie, which shows I was not a self-taught engineer, as some of my friends have supposed, and stated, on what grounds I know not.[10]

Bell clearly viewed his work with Rennie as part of his professional education and not simply as a job. Unfortunately there do not appear to be any references to Bell in Rennie's voluminous papers, so it is impossible to say exactly what work Bell did for Rennie or how he even secured the position. One may perhaps speculate that there was some judicious use of a Scottish millwright network, perhaps involving a Bell family acquantaince with Rennie's first master, Andrew Meikle.

However the place was secured Bell was now at the very

centre of events in a way that he had not been before. Bo'ness and Bellshill might have been all very well in their practical workaday way but London was where contacts were to be made and where ideas most rapidly circulated. For example while Bell was in Rennie's employ, Rennie was corresponding with Lord Stanhope on the cost of applying a Boulton and Watt steam engine to Stanhope's patent method of propelling ships.

John Rennie had gone south in 1784 and after paying the almost obligatory visit to the Soho, Birmingham, works of Boulton and Watt—one of the technological wonders of the age— he went on to London to work initially for Watt and then later to set up in business on his own account. He was to build up a large and varied engineering practice which would include canal construction, drainage in the Lincolnshire fen-country, the construction of bridges, docks and harbours, and a wide consulting practice.

His first large contract in London had been the Albion Flour Mills in Blackfriars. This was a major undertaking and some of the elements in it would provide excellent guidance for work Bell was to carry out, plan or just dream of in later life. The Mills were to be powered by Boulton & Watt steam engines and the construction of the mill buildings was an architectural and civil engineering endeavour of the largest scale. Rennie's part in the project was the erection of the engines and the construction of the mill machinery. The second engine was laid down early in 1789 and this operation doubtless afforded the young Bell many valuable insights into many aspects of engineering, not least the use of steam.

Rennie, despite his very varied practice, was to be little concerned with the application of steam to ships, and presumably in this followed Watt, who as we have seen warned Bell against involvement in this seemingly futile activity. Rennie corresponded with Stanhope and helped assess the worth of Robert Fulton's torpedo but does not himself seem to have tried any experimental work in this field. Bell presumably can thus have gained little that was to be directly relevant to his steamboat aspirations from his time in London with Rennie. On the other hand the varied experience he gained in the metropolis, and perhaps more importantly the prestige of having worked for Rennie, can have done nothing but good to his prospects and his ability to attract support for his later schemes.

CHAPTER 3
"Wright in Gorbals"

Around the year 1790, his travels in search of training and experience over, Henry Bell returned home to Scotland and in 1790 or 1791, in partnership with one James Paterson, set up in business in the Glasgow area as a builder, house carpenter or wright.

Who James Paterson was, and how long the partnership lasted, are both uncertain. Jones' *Glasgow Directory for 1789* [1] lists a James Paterson, collector (or Treasurer) of the Incorporation of Wrights, resident on the north side of the Bridgegate. However there are good reasons to suspect that this is not the Paterson with whom Bell was in partnership. Firstly, the fact that this Paterson was well-established as evidenced by his having risen to a senior position in the Incorporation would make it improbable that he would be available or likely to go into partnership with an unknown incomer, even one with Bell's varied work experience and impressive contacts. Secondly, and more importantly, is the difficulty over the Incorporation itself.

Strictly speaking, if Bell was going into business as a carpenter or housebuilder in Glasgow he would have been required to join the appropriate trade guild, the Incorporation of Wrights. The trade incorporations were very powerful and important organisations of master craftsmen and employers and must not be confused with trade unions. They were indeed an essential and integral part of the city's government and had automatic representation on the City Council as well as an important regulatory function in establishing the conditions, and controlling training, entry and numbers in the various trades. Bell was in due course to join the Incorporation of Wrights, but not until much later, in October 1797. It seems very unlikely that an official of the Incorporation could have countenanced being in partnership with an incomer who had for so long failed to join the Incorporation. Bell, certainly in 1794, and possibly for most of the 1790s was resident in the Gorbals, then a separate administrative entity and as such outwith the control of the City Incorporations. While his location in the Gorbals explains the delay in joining the Incorporation it also makes it rather unlikely that he was in partnership with James Paterson of the Bridgegate.

Bell was duly enrolled as a burgess and guild brother of the City of Glasgow, by purchase, on 17th October 1797 and was admitted three days later to the Incorporation of Wrights.[2] By 1799, the first year in which he can he be traced in local directories, he had become well established in the city. Well enough established, at any rate, to be able to describe himself as an architect. Such a description could at that time, of course, be self-applied without any specific or obligatory training or qualification. His home at this stage was listed as "corner of King Street, East Side." Two years later he had moved to Buchanan Court, Trongate[3], and by 1804[4] was resident at Melville Place, 132 Trongate. There is indeed a possibility that this property was built by Bell. Edward Morris in his biography states that Bell built a property in Glasgow called Melville Court.[5]

No such address can be traced at this period and it seems more than likely that this is a mistake for Melville Place. Bell's final permanent address in Glasgow was 26 Ingram Street, this address appearing in the 1807 directory. This edition is the last in which he is listed and would tend to confirm the timing of a move to Helensburgh and the building of the Baths Inn at this period. Indeed in July and September 1806 Bell received accounts from Scotts of Greenock, the shipbuilders and engineers, addressed to him at Helensburgh and it would seem very likely that he was at these times resident in Helensburgh supervising the construction of the Baths Inn.[6]

Bell's residential movements in Glasgow, as far as they are recorded, do certainly suggest a successful career and are indeed indicative of a degree of upward social mobility. Ingram Street was part of the westward expansion of the city and had only been opened up in 1781. It housed the splendid Assembly Rooms designed by Robert and James Adam and built in 1797. These were to be extended by Henry Holland in 1807, at the time Bell was living in the street. Ingram Street was a considerably more desirable address than King Street, which although only a few hundred yards away, was in a more industrial area—Glasgow was then a very small and compact city with contrasting areas of very different character and social composition cheek by jowl.

Whatever uncertainties may exist about Bell's Glasgow period, one definite fact is that in 1794 he married Margaret Young. The records for the City Parish of Glasgow record this wedding on 26th March 1794 as being between Henry Bell, Wright in Gorbals and Margaret Young, Residenter in Glasgow. The banns were apparently published in Bell's parish on 12th March and in Margeret's parish on 23rd March. Despite this dutiful recording of the wedding in the records of the Established Church it is believed that the couple were in fact married in the Anderston Seccession Church, Margaret Young being attached to this denomination.

Margaret was the daughter of Robert Young and Janet Hamilton and she was born in the village of Kilbride in the Lanarkshire parish of Lesmahagow in 1770. Robert Young must presumably have died while Margaret was still quite a young girl because in June 1783 her mother, Janet, re-married. Margaret's step-father was to be Thomas Dykes of Burnhouse, a resident in the Renfrewshire parish of Eaglesham. The "of Burnhouse" appendage to his name suggests that Dykes was a landowner, however small his property may have been. Burnhouse is a farm, near to the tiny village of Auldhouse, some 600 feet above sea-level on the boundary of Renfrewshire and Lanarkshire. Even today the area is an isolated one, despite being only three or four miles from the rapidly expanding new town of East Kilbride. The suggestion of a certain degree of affluence in the Dykes family is born out by the Parish Record description of Margaret as "Residenter in Glasgow". The absence of any stated occupation is suggestive of independent means.

Margaret was by repute something of a beauty and her obituary remarks:

> ... till the very verge of her long life the evidence of her early beauty and vivacity were strongly marked.[7]

Perhaps more to the point the writer notes that:

> Mrs Bell was a woman of strong practical good sense—a quality not generally looked for in country belles—and there can be no doubt but that [it] was exercised beneficially in promoting her husband's efforts, as well as her cheerful good nature in sustaining him amid many discouragements. To the latest period of her life she always spoke of her husband and his plans, not only in such a way as to show her intimate acquaintance with them, but with such a degree of respectful regard to even the most trivial of them, that one could not fail to feel impressed with the deep sense of honour and love in which she seemed to cherish his memory.

This suggestion of Margaret's support for Henry Bell and the pride she took in his work is confirmed by an article written by James Barr for a Dumbarton paper in the 1890s and recalling a visit he and a friend had paid to the Baths Inn around the early 1840s:

> On entering we found her [Mrs Bell] to be a quiet, unpretending little woman, modestly attired in black. She recognised James [Barr's companion] at once, and the cordiality of the greeting was to me sufficient proof that he and her husband had really been intimate friends. Before we left she

showed us models of different kinds which he had made from time to time, and the saddened expression of her face when referring to him proved that her bereavement had been deeply felt.[8]

Edward Morris noted that Bell was for many years lame and infirm and that:

> . . . in consequence of this, Mrs Bell had in a great measure to manage his affairs, and wisely and prudently did she discharge these double duties—as all will testify who have the pleasure of knowing her.[9]

While Bell can hardly have been the easiest man to be married to, it would seem from all accounts that Margaret was a more than equal partner in the relationship and one cannot help but feel that Bell's success, however limited it may have been in financial terms, owes much to Margeret's sound work in keeping the Inn running smoothly and Henry out of trouble.

More than this, the marriage may well have been a sound financial arrangement. The evidently increasing prosperity of Henry Bell in the late 1790s and early 1800s may possibly owe something to a good marriage settlement. Certainly his connection with the Dykes family was to be useful many years later when three members of the family, Thomas, John and Janet, presumably Margaret's stepbrothers and step sister, were to buy shares in the *Comet*.[10]

Margaret, when her marriage to Henry was recorded, was identified in the Parish Register as a "Rcsidenter". There is however a suggestion that she was, either when she married Bell or possibly earlier had been, housekeeper to Archibald Newbigging of Mill-Bank. Newbigging was a Baillie or city magistrate in 1813 and 1814 and had long been a leading figure in city life, having served as River Baillie in 1796, as Convener of the Trades House, the joint body of the Trade Incorporations, in 1799/1800; and as City Treasurer in 1806. The source for this suggested link between Margaret Bell and Newbigging is John Robertson, the engine-builder of the *Comet*. John Buchanan notes that in 1853 he was told by Robertson that:

> Subsequently Bell took up his quarters at the Baths, Helensburgh, which was built, he believes by Baillie Newbigging of Glasgow, with whom Mrs Bell had been housekeeper, and this led to Bell going to Helensburgh at all.[11]

The other evidence, such as the Register of Sasines, suggests that Robertson was not correct about Newbigging building the Baths Inn, although Newbigging certainly took a financial interest in it some years after it was built. Bell was in fact to mortgage it to "Archibald Newbigging, Merchant in Glasgow", for £2000 in May 1810.[12] The Baths Inn featured in a number of other financial transactions involving Newbigging and it was, in 1819, disposed of to the Trustees on the sequestrated estate of Archibald Newbigging and Co Merchants and Linen Printers.[13]

It is perfectly possible that Robertson, a close acquaintance of Bell's and working in the small and close-knit business community of Glasgow, knew something of all these transactions and had understandably assumed that Newbigging had been involved in the Baths Inn before 1810, that is in the period of its building around 1807. However Robertson's apparent error in this respect (and it must be recollected that he was an old man being interviewed some forty years after the events) does not in itself invalidate the statement that Margaret Bell had been Archibald Newbigging's housekeeper. The obvious possibility exists that this link had facilitated the dealings between Bell and Newbigging in 1810. Indeed there is a chance that there might have been some kinship between Margaret and Newbigging—it being by no means an uncommon practice to employ a relative in the trusted position of housekeeper. It is of course also not outwith the bounds of possibility that Bell may have had previous business dealings with Newbigging or his firm. Their trade as linen printers could well have provided some opportunities for Bell in his role as millwright, engineer or architect. This indeed would seem to have been the type of work he was known for. Interestingly enough one of the few contracts we know of in Bell's architectural or engineering career is from another textile firm, the Messrs Kibble of Dalmonach, Dumbartonshire, who employed him to rebuild their mill in the Vale of Leven in 1812.

Little is recorded, and not much can be discovered about Bell's activities in the years between his return to Scotland and his move down to Helensburgh. His only positively documented work as an architect seems to be the contract he won in 1799 to provide plans for a new church for the Lanarkshire parish of Carluke.[14] The old pre-Reformation church had proved to be too small for the needs of the rapidly growing population of the parish. The Heritors, the chief landowners of the area who had the responsibility for the provision of church buildings, were instructed by the Presbytery of Lanark to make better provision. They resolved to build a new church and advertised in the Glasgow press for plans. Bell's submission met with approval, after the removal of "all superfluous ornament", and he was rewarded with a fee of two and a half guineas—£2.12.6. The contract to build the church went to another Glasgow wright, James Kay.

On the evidence of his changes of house Bell would seem to have built up a reasonable business in and around the city and the few glimpses we have of him in this period confirm this idea. In his own autobiographical notes, written to assist Edward Morris, he says, talking of his training under Inglis and Rennie:

> In all the above occupations, I made it my study to turn them
> to some practical use, of which numerous public works in
> Glasgow bear witness.[15]

"Public works" in this context is a reference to factories or mills, rather than to civic undertakings. Morris later says that Bell's partnership with Paterson lasted from 1791 to 1798:

65

> ... during which period they finished many public works in and around Glasgow.[16]

and also states that Bell built a large flour mill at Partick. This is presumably the Clayslap Mill, extended by the City Corporation in 1801 and according to the contemporary Glasgow historian, and defender of Bell, James Cleland:

> ... not inferior to any in the Empire, in point of situation, management, and the internal arrangement of the machinery. The principal mill has four floors; is 207 feet long, and 41 feet wide. . . The value of the whole may be estimated at something between £45,000 and £50,000.[17]

If Bell was indeed associated with this major project then he must have been both well-established and well thought of to secure such an important civic contract.

David Napier, who made the boiler for the *Comet* wrote in his manuscript memoir that Bell:

> ... was very frequently in the foundry getting castings for buildings he was superintending.[18]

And he also had, presumably, similar work carried out at other firms such as Scotts.

There is a tantalising reference to Bell's work at this time which appeared in the report of a House of Commons Select Committee in 1822. This in the course of discussing the history of steam navigation goes on to note:

> Mr. Bell continued to turn his talents to the improving of Steam apparatus, and its application in various manufactures about Glasgow. . .[19]

Sadly nothing more precise is said but the implication of Bell working with steam engines in mills and factories in the Glasgow area is quite clear. Certainly there were a fair number of steam engines in place at this time. Even in a small country town like Dumbarton the local engineering and blacksmithing firm of James and John Napier boasted two steam engines in the early 1800s.

These indications of significant contracts and technically advanced work are somewhat at odds with the account of Bell's life appearing in Chambers' *Edinburgh Journal* in 1839. It describes this period of Bell's career in the following rather dismissive terms:

> About 1790, he settled in Glasgow, where he wished to be employed as a constructor of public works; but limited resources were a sufficient bar to his advancement in such a course, even if he had possessed all the requisite personal qualifications. He therefore remained for several years in the obscure situation of a house-carpenter—known, if known at all, only for a disposition to mechanical scheming. . [20]

Admittedly a two and a half guinea contract for Carluke church hardly counts as evidence of a successful career, but it is perhaps significant that Bell felt confident enough to submit plans and that they won approval. It seems fair to conclude that other work was done but has not been recorded.

One interesting aspect of Bell's Glasgow period, the evidence for which has survived, is his elaborate proposal on the subject of a water supply scheme for the city.

Prior to 1804, Glasgow's only water supply had come, inadequately and uncertainly, from wells. In that year William Harley built a reservoir in Nile Street, which was supplied from his lands of Willow-bank and two years later, in 1806, a committee was formed to raise a subscription for supplying the city with filtered water from the Clyde. This committee was incorporated as the Glasgow Water-Works Company. An Act of Parliament was obtained and Thomas Telford commissioned to produce plans for bringing river water into the city from Dalmarnock. A filter works and pumping station were to be built at Dalmarnock, a water tower and basin in Sidney Street, Gallowgate, and a reservoir in the upper part of town at Rottenrow.

These proposals did not go unnoticed by Henry Bell, who in March 1806 published a broadsheet addressed *To the committee and other subscribers interested in supplying the City of Glasgow with water*.[21] In this he criticises, comprehensively but tactfully, Telford's plan:

> It is far from my intention to dispute the abilities of the Gentleman who was lately employed to give a general survey and estimate of the expence.

He particularly questions the proposal by Telford to draw water from the River Clyde near the city on the quite reasonable, and indeed far-seeing, grounds of the danger of pollution from industry. He also criticised the need, unavoidable in Telford's plan, for an engine to raise water to higher parts of city. His argument on this point was based on the contention that such engines would cause a nuisance and damage to property values and that the consumption of coal by the engines will raise the price of fuel in the city; it is perhaps possible to feel that he was scraping the bottom of the barrel for arguments at this point. He also queried the cost of fuel and the cost of maintaining the engines and declared that the danger of breakdown in the engines and forcing pipes could leave the city without water.

Bell offered the Committee an ambitious alternative scheme. He proposed to draw his supply of water from the Clyde at Stonebyres Linn, some thirty miles up river near Lanark, at a point 370 feet above sea level. This supply, he claimed, was high enough to ensure a secure gravity supply to all parts of the city. The water was to be carried by a 6'x3' canal or cut, through filters, to a reservoir above the city from whence it would flow down to small supply reservoirs.

Bell stated that he had made a survey of the area and printed an analysis of his estimated costs which totalled £23,500. This covered his planned reservoir, associated works such as bridges and cottages and the thirty mile long canal—the last two miles of which were to be covered over to prevent soot and pollution from affecting the purity of the water. With a view to fending off possible objections by riparian proprietors he suggested that if complaints were raised to the amount of water being drawn off at Stonebyres Linn then it would be possible to create a reservoir, near the higher parts of the river, to provide compensation to the river, in time of drought, equal to two or three months consumption for the city.

In conclusion he says that he has not yet carried through a survey into the city and is thus unable to give details of the small reservoirs and pipes that will be necessary. However he notes that as Stonebyres Linn is considerably higher than any part of the city the natural pressure of flow will require pipes of a smaller size than any scheme based on a reservoir situated at a lower level.

Whatever its merits may have been, Henry Bell's scheme was not adopted, and the Company proceeded with Thomas Telford's proposals—not without some difficulties being experienced as the filter beds tended to fill up. Later on this scheme was supplemented by water which was drawn from other side of the river at Farme, near Rutherglen. In this case the supply of water was filtered through natural sand and gravel beds and pumped by steam engines through a flexible pipe laid under the bed of the river and fed into the Dalmarnock mains created under the Telford plan. When such measures proved to be inadequate to meet the needs of the growing city the corporation in the middle of the nineteenth century adopted a scheme similar to that proposed by Bell, although in the opposite direction, north-west to Loch Katrine in the Trossachs, rather than south-east to Lanark.

The fact that Bell had worked out such a scheme and had gone to the trouble of preparing and publishing what amounts to a prospectus for an alternative to the recommendations of the eminent Thomas Telford may, one might think, tend to confirm the Chambers' *Edinburgh Journal* comment about "... a disposition to mechanical scheming ... ", but, equally, it does also suggest a background, a status, and a public persona somewhat higher than "... the obscure situation of a house-carpenter. . . "

Bell's means were, it would seem, not totally confined to the earnings from his business, nor should it be thought that he came from a background of poverty. The Bell family was numerous and prosperous, and his marriage had perhaps brought a useful dowry. After his period of residence in Glasgow Bell owned land there—a piece of ground of 555 square yards—which had come to him in 1813 from Robert Easton, farmer of Woodcockdale in West Lothian.[22] Robert Easton was presumably a relative of Bell's mother, Margaret Easton, who after her husband's death had moved to Woodcockdale and died there in 1802.

The conclusion one is obliged to draw from such evidence as is available is that Bell, at this time, was a reasonably successful contractor and financially secure. A man who was well known to, and respected by, his contemporaries in the city such as David Napier and John Robertson. The fact that both these men willingly accepted orders from Bell for the *Comet* suggests that he was considered a good risk for the sums involved. This surely gives us some indication of his financial status and taken with what must have been his major investment in the Baths Inn again argues that Bell was more than just a house-carpenter with ambitions, interests and pretensions above his station in life. Undoubtedly the move into steamships was an expensive investment and one that strained his finances for the rest of his life. It may well be that some commentators, knowing of Bell's later financial difficulties, which were indeed very public, have assumed that the Glasgow period was equally fraught with problems.

The tolerance which Napier and Robertson extended to Bell over his failure to meet their bills, and the unstinted praise given to his pioneering efforts by knowledgeable contemporaries such as James Cook and the other leading engineers of Glasgow would also incline one to believe that Bell must have been both a man endowed with a considerable and attractive personality and equally a person of recognised ability and not just, as is sometimes suggested, the self-taught bungling experimenter who unexpectedly succeeded.

Although John Robertson was to report that Bell was:

> . . . restless, unmethodical and by no means a good hand at machinery[23]

it is somehow hard to envisage Robertson, Napier and John Wood all blithely accepting contracts from Bell, with all the possible dangers that might follow, if he were so bad a risk as is sometimes suggested. If these other Clydeside engineers willingly dealt with Bell one is obliged to question the full truth of the frequently repeated tale that, in the words quoted in Thomson's *Biographical Dictionary of Eminent Scotsmen* he had:

> . . . many of the features of the enthusiastic projector—never calculated means to ends, or looked much further than the first stages or movements of any scheme. His mind was a chaos of extraordinary projects, the most of which, from his want of accurate scientific calculation, he never could carry into practice.[24]

A fairer assessment would certainly take into account his lack of method and his evident over-enthusiasm but would balance this with an acknowledgement that he not only appeared to be able to run a successful business in Glasgow but was also able to devote time and imagination to solving a problem that had eluded better trained and better funded experimenters all round the world for many years. The list of artists, engineers, philosophers, savants and amateur experimenters who had attempted to make a success of mechanical and steam propulsion is long and distinguished. The fact that the first person to make a practical success of the endeavour in Britain was the enthusiastic, restless and unmethodical Henry Bell does seem at times to be almost a cause of offence to some contemporary and to some later writers—possibly because of the apparent social and educational gulf between the Torphichen stone-mason and the learned men of science who had laboured so long and so fruitlessly in this field.

Chapter 4
"... to answer the end": The Problem of Steam Navigation

The search for a means of propulsion for ships, other than wind or oar, is a very ancient one. The idea of using a paddle would seem to have originated many centuries before the steam engine was available to provide a suitable source of motive power. Ships propelled by paddle wheels driven by oxen are supposed to have been used by the Romans. Perhaps inevitably, the Chinese also were to the fore in this area of activity and man-powered paddle-driven ships are claimed to have been used by them in the 7th century AD.

Leaving aside the somewhat dubious technique of ox-power, the attraction of the man-powered paddle over the oar were two fold. While the oar is a highly efficient means of converting human energy into propulsive force it requires a degree of skill—a crank on the contrary requires nothing but brute strength. It is also difficult to use more than one oarsman on each oar but a number of workers can be combined on a crank thus multiplying the force applied.

Mechanical propulsion by paddle wheels was, in effect, re-invented time after time, by scientists, engineers and inventors. These included, again perhaps inevitably, Leonardo da Vinci, who depicted a variety of means of achieving paddle-wheel propulsion. This wasteful replication of the inventive process provoked the magisterial authority of the *Encyclopedia Britannica* in its survey of steam navigation in the 1824 Supplement to observe:

> Such waste of ingenuity is really a serious evil, arising from inattention to the history of the mechanical arts, and the want of a public repository to exhibit their successive improvements.[1]

Some writers, including Bell's biographer Edward Morris, claim that the first attempt at a steam-powered vessel was made in 1543 by the Spaniard Don Blasco de Garay. This rather remarkable claim, which of course pre-dates by 180 years Newcomen's invention of the atmospheric steam-engine, has been exploded by H Philip Spratt in his *The Birth of the Steamboat* who points out that the claim that de Garay's experiment involved steam propulsion was not made until 1825. He argues that this suggestion arose from a confusion between a cauldron of hot water for defensive purposes which was installed on the deck of the experimental ship the *Trinidad* and a steam boiler. Spratt further points out that while the ship was fitted with paddle wheels records show that these were each powered by twenty-five men.

A variety of ingenious, if impracticable, schemes exercised the minds of many workers in this area. However not until the steam engine became available was there a power source which would make mechanical propulsion a reasonable proposition. The coming of the Newcomen engine was the key to the widespread 18th century interest in the topic. The *Encyclopedia Britannica* commented on the atmospheric engine:

> ... it does not seem to have required any great stretch of imagination to direct such an efficient power to other purposes besides the raising of water.[2]

Jonathan Hulls, a Gloucestershire inventor, born in 1699, is credited with the first theoretical attempt to apply the Newcomen engine to navigation. Hulls obtained a Patent on 21st December 1736 and described his plans in a pamphlet published in the following year, *A Description and Draught of a new invented Machine for carrying Vessels or Ships out of or into any Harbour, Port or River, against Wind or Tide, or in a Calm*. Hull's proposals involved the placing of a Newcomen engine in a tow-boat and he would seem, in theory at least, to have produced one solution to the problem of converting the rectilineal motion of a piston rod into the rotary motion of the paddle.

The Newcomen engine was of course a single-acting cylinder and the Hull engine lifted, on the effective down-stroke, a weight which, on the engine's return stroke, would provide by its descent the effect of a double action engine. This process was converted into continuous action by means of ratchet wheels driven by the down-stroke and the weight.

Authorities disagree on whether Hull's scheme was ever subject to practical experiment—Spratt declares that due to the withdrawal of support from his patron no trials were carried out, while the *Dictionary of National Biography* mentions trials on the River Avon at Evesham in Warwickshire in 1737. In any event no practical working vessel resulted and Hull turned his attention to other schemes, such as a machine for weighing coins.

The French were particularly active in the development of the steamboat—Denis Papin (1647-1714) had from 1690 written and experimented on paddle-boats. In the 1770s a series of French experimenters had worked on the problem and in 1783 the Marquis de Jouffroy D'Abbans succeeded in getting his second steamship the *Pyroscaphe* to sail upstream on the River Saone. The Revolution made further progress difficult for the Marquis although after the Restoration of the Bourbons he was to build another steamship, tactfully named *Charles-Philippe* after the Comte D'Artois, later Louis X of France, which sailed on the Seine in 1816.

In America too, a variety of theoretical and practical work had been carried out. The political pamphleteer Thomas Paine advocated the application of steam to navigation and John Fitch (1743-1798) successfully demonstrated a steam-driven boat on the Delaware River in 1786.

Britain, as the birthplace of the Industrial Revolution, was as one would imagine, not behind-hand in the search for mechanical propulsion. Patrick Miller of Dalswinton (1731-1815), was a Dumfriesshire landowner who had made a considerable fortune as a banker. He had a long standing interest in maritime matters and had developed the carronade—a short range naval cannon named after its first maker, the Carron Company of Falkirk. Miller's cannon proved to be highly successful and weapons of this type were widely used by all powers in the period of the Revolutionary and Napoleonic Wars. Miller had conducted experiments with multiple-hulled vessels and took out a patent in 1796 for a "vessel which will draw less water than any other of the same dimensions; and which cannot founder at sea". Working in concert with James Taylor, a graduate of Edinburgh University, whom he had employed as tutor to his sons, Miller then turned his attention to mechanical propulsion. At first this simply was to take the form of hand-cranked paddle wheels. Taylor was later to describe these initial experiments:

> In the summer of 1786, I attended him repeatedly in his experiments at Leith, which I then viewed as parties of pleasure and amusement. But in the Spring of 1787 a circumstance occurred which gave me a different opinion. Mr Miller had engaged in a sailing match, with some gentlemen at Leith, against a Custom House boat (a wherry) which was reckoned a first-rate sailer.[3]

The race was run in the Spring of 1787 on the Firth of Forth with Miller's double-hulled vessel propelled by four men proving victorious. Taylor noted:

> Being then young and stout, I took my share of the labour of the wheels, which I found very severe exercise; but it satisfied

(top) Jonathan Hull's steamboat – the first attempt to apply the Newcomen engine to steam navigation
  (Mitchell Library, Wotherspoon Collection)
  (bottom) The Dalswinton Steamboat
  (Science Museum)

me that a proper power only was wanting to produce much utility from the invention.

Taylor claimed that Miller's initial motive in experimenting in mechanical propulsion was to produce no more than an emergency source of propulsion for vessels in distress but that he persuaded his employer of the utility of steam propulsion, if only for inland navigation.

William Symington, the engineer of the *Charlotte Dundas*, claims to have been on board this experimental vessel on one of its voyages and expresses a view on the impossibility of a man-powered paddle-boat being successful and confirms Taylor's claim that he mentioned the application of steam-power to his employer.

> The truth is that Mr Miller's experiments were unsuccessful in the eyes of everyone but himself, as many can declare who are still living (myself among others,) and who laboured in the turning of the wheels. I can well sympathize with Mr Taylor, the tutor of his son, who exclaimed in the agony of his exertions, that he wished for a steam engine to assist them.
>
> The hint of the tutor, it would appear from Mr Miller's own narrative, is the first suggestion of the practicability of applying the steam engine towards the end he had in view; and the public will be better able to appreciate the effect of the admission when they are informed, that I was on board of the same boat, or at least I was on board on one occasion when the necessity of such assistance was generally admitted.[4]

Symington's statement and claim to first hand knowledge is particularly interesting because Taylor and Miller Jnr suggest that Symington was recruited when Patrick Miller decided to conduct his experiment on Dalswinton Loch and as we shall see both tend to downgrade Symington to an operative, merely executing the ideas of Miller or Taylor. If, however, Symington was working with Miller in the summer of 1786 or the spring of

1787 then it does cast a slightly different light on Taylor's version of events.

In the summer of 1787 Miller wrote up and circulated an account of his experiments in man-powered paddle-ships and included, at Taylor's urging, a mention of the possibilities of steam power. Later Miller and Taylor went to Edinburgh and, in the version given by Taylor, made contact with William Symington.

> At this time William Symington, a young man employed at the lead mines at Wanlockhead, had invented a new construction of the Steam Engine by throwing off the air pump. I had seen a model work and was pleased with it, and thought it very answerable for Mr Miller's purpose. . . Being acquainted with him, I informed him of Mr Miller's intentions and mine, and asked if he could undertake to apply his Engine to Mr Miller's vessels. . .[5]

Symington had in fact produced a practical steam engine which avoided the Watt patent on the separate condenser. Symington agreed to work on Patrick Miller's scheme and it was arranged that the experiment should be tried out on a boat on Dalswinton Loch on Miller's Dumfriesshire estate. This experiment took place on 14th October 1788 and in Taylor's words:

> . . . a more complete successful & beautiful experiment was never made by any man, at any time, either in art or science. The vessel moved delightfully, and notwithstanding the smallness of the cylinder (4" dia.) at the rate of five miles an hour! After amusing ourselves a few days, the Engine was removed, and carried into the house, where it remained as a piece of ornamental furniture for some years.

Despite the somewhat dilettante approach to scientific progress revealed by Miller's abandonment of the Dalswinton Steamboat and the conversion of the engine into a piece of furniture Taylor was authorised to write up an account of the work for the local press and an account in the following terms appeared in the *Scots Magazine* for November 1788:

> On October 14, a boat was put in motion by a steam engine, upon Mr Miller of Dalswinton's piece of water at that place. That gentleman's improvements in naval affairs are well known to the public. For some time past, his attention has been turned to the application of the steam engine to the purposes of navigation. He has now accomplished, and evidently shown to the world, the practicability of this, by executing it upon a small scale. A vessell, twenty-five feet long, and seven broad, was, on the above date, driven with two wheels by a small engine. It answered Mr Miller's expectations fully, and afforded great pleasure to the spectators. The success of this experiment is no small accession to the public. Its utility in canals, and all inland navigation, points it out to be of the greatest advantage, not only to this island, but to many other nations in the world. The engine used is Mr Symington's new patent engine.[6]

Taylor and Miller discussed matters further and agreed to conduct additional experiments, this time on a larger scale, on the Forth and Clyde Canal. It must be said that Taylor's account of the Miller/Taylor/Symington steamboats was written to persuade the government that his was the mind behind the project and that any reward for the invention should go not to Bell but to himself. For this reason the Taylor version must be taken with some reserve and the account of Taylor always urging on a somewhat reluctant Miller may be considered slightly suspect.

In the Spring of 1789 Taylor went to the Carron Company works at Falkirk with Symington to supervise the construction of an engine. In December of that year successful trials were run on the Canal under the supervision of Taylor and Symington. Taylor reported his success to his employer but Miller seemed to be growing less interested in the subject. To be more exact Miller was becoming more interested in, and more financially committed to, agricultural improvements on his estates. He was also angered at the substantial cost over-runs against Symington's original estimates and resolved to have nothing more to do with engines. The experimental vessel was laid up and the engines removed and returned to the Carron works.

Taylor was however instructed to submit an account of the successful experiments to the press and this was drawn up in consultation with an Edinburgh advocate, Robert Cullen. This procedure was doubtless followed through to ensure that the claim to intellectual and practical priority, which would be of some significance if and when the experimenters sought a patent or commenced the commercial exploitation of their work, was expressed in a proper and legally watertight manner. The *Caledonian Mercury* of Saturday 20th February 1790 included the following extract from a letter dated at Falkirk on February 12th 1790:

It is with great pleasure that I inform you, that the experiment, which some time ago was made here upon the great canal, by Mr Miller of Dalswinton, for ascertaining the power of the Steam Engine when applied to sailing, has lately been repeated with very great success. Although these experiments have been conducted under a variety of disadvantages, having been made with a vessel built formerly for a different purpose, yet the velocity required was no less than from six and a half to seven miles an hour. This sufficiently shows that, with vessels properly constructed, a velocity of eight, nine or even ten miles an hour may be easily accomplished. The advantage of so great a velocity in rivers, straits, etc, and in cases of emergency, will be sufficiently apparent, as there can be few winds, tides, or currents, which can easily impede or resist it; and it will easily be evident that, from a slower motion, the utmost advantages must result to inland navigation.

The *Mercury* commented:

> It is with the greatest satisfaction that we have received this intelligence from our obliging correspondent. Every well-wisher to the extension of arts and of commerce must be highly gratified with the signal success of this important experiment, which bids fair to introduce an improvement, which, by greatly facilitating and rendering more easy and speedy the intercourse by means of navigation, must not only be highly advantageous to our own country in particular, but to the commerce of the world at large, and to mankind in general.[7]

So matters stood for some years as far as Taylor was concerned. He left the service of Patrick Miller, although remaining on sufficiently good terms with his former employer as to be able to call on him socially in Dumfriesshire. On his visits to Dalswinton in the 1790s Taylor found the Laird preoccupied with agricultural improvements and unwilling to proceed further with steamship experiments. Taylor, as one may imagine from his employment as a tutor, did not have the financial resources of his former employer which would allow him to finance further experimental work himself.

Elsewhere in Britain work was also going forward. An ill-documented attempt at a man-powered paddle ship was reported to have been made at Leith by a Mr Laurie in 1787 and the Earl Stanhope obtained a patent in 1790 for a steam propulsion system driving paddles which operated like a duck's webbed foot. In March 1793 Stanhope's experimental vessel *Kent* was launched on the Thames using a 12 hp engine. He had tried to utilise an engine with the Boulton and Watt separate condenser but James Watt, who would seem to have viewed the application of steam to marine navigation with a high degree of scepticism, appears to have refused his cooperation. Stanhope had, around the time of winning his patent, consulted John Rennie who had provided him with figures for the cost of applying a Boulton and Watt engine to his vessel. Stanhope was later to correspond with Robert Fulton who recommended to him the more conventional form of paddle.

In 1800 William Symington was commissioned by Lord Dundas, the Governor of the Forth and Clyde Canal Company, to fit up an experimental steamboat for service on the canal as a tracking or tug boat. This vessel, which was normally, if seemingly unofficially, called the *Charlotte Dundas*, after Lord Dundas's daughter, was built at Grangemouth by Alexander Hart. Hart was the son of the noted Bo'ness shipbuilder and sometime employer of Henry Bell, George Hart. This vessel was fitted with William Symington's patent engine and ran trials from June 1801 with some success. However, the Canal Company became increasingly unhappy about the steamboat project and in January 1802 the Scottish committee of the Company minuted that:

> The Committee observing by the books of the Company that the steam boat constructed by Mr Symington has already cost £858.12.1 and as several of the gentlemen of the committee have good reason to think that she will by no means answer the purpose of tracking vessels, it is the opinion of this meeting that Mr Symington should be desired by the Governor to give in a statement of all the apparatus belonging to the boat and account of his whole expenses, so that a final settlement may be made with Mr Symington and steps taken to turn the boat and apparatus to best use.[8]

Although Lord Dundas still had confidence in the principle of steam power and introduced Symington to the Duke of Bridgewater, the great English canal promoter, the Canal Company finally decided that the *Charlotte Dundas* was not going to answer their purpose. In any case they were concerned about the damage to the canal banks caused by the wash from the paddle wheels and ordered her to be converted to a ballast boat.

One of the problems of the Forth and Clyde Canal Company was a split between the main board, meeting in London under the Governor, Lord Dundas, and the Scottish Committee, meeting in Glasgow. The Glasgow Committee was unsympathetic to Symington's experiments but despite this Dundas authorised Symington to proceed with the building of a second experimental vessel. A great deal of confusion has arisen over these two vessels which are conventionally referred to, though contemporary records do not support this, as *Charlotte Dundas I* and *II*.

The history of these two vessels is complicated and has been comprehensively dealt with by Harvey and Downs-Rose in *William Symington: Inventor and Engine Builder*. For the present purpose it is sufficient to note that *Charlotte Dundas II* made its first trial in January 1803 on the Glasgow stretch of the Canal and performed tolerably well, though not apparently at a speed sufficient to satisfy Symington. In a later trial she towed two cargo vessels from Lock 20 at Wyndford to Port Dundas in Glasgow—a distance of eighteen and a half miles in nine and a quarter hours, a demonstration which showed that *Charlotte Dundas* was capable of useful work, but which was no better from the point of speed than the time taken by the conventional horse tracking.

The Canal Company was not impressed and having rejected the first vessel, members were reluctant to change their minds. Dundas was at this juncture unwilling, or possibly unable, to press the issue and the unfortunate Symington was faced with a long delay before the Company met his costs. Doubly unfortunate, Symington, who had been commissioned by Bridgewater to build eight steam tugs for his English canals, found that this order was cancelled when the Duke died in March 1803.

The unwanted *Charlotte Dundas II* was laid up on the Canal where it became an object of curiosity and was apparently visited by a number of people interested in steam navigation, including Robert Fulton and Henry Bell. James Taylor, writing in 1825, stated that

> Whilst the Boat of the Canal company lay at Lock 16, Mr Fulton, the American Engineer, was travelling for information in the line of his profession, and while he and a Mr Bell of Glasgow were visiting the Carron works, they were informed of the Steam Boat. This was exactly an object for

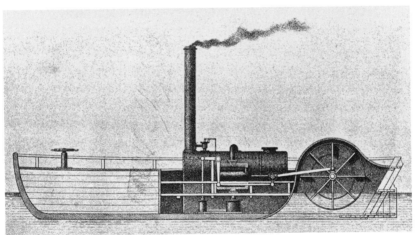

(top) William Symington in old age.
(bottom) Plan of the Charlotte Dundas, *built at Grangemouth*
for use on the Forth and Clyde Canal
(Dumbarton District Libraries)

Mr Fulton's observation, they accordingly called on Mr Symington at Falkirk, and requested to be shown the Steam Boat. This was readily granted, and a full description given. It appears that these gentlemen had afterwards corresponded on the subject and when Mr Bell was at any loss, he paid visit after visit to Lock 16 to get over his difficulty. It appears also that both these gentlemen had projected a scheme of turning the invention to a speculation of profit on their own account. In consequence Mr Fulton brought forward his first Boat upon the Hudson in 1807, and he and his country vainly claimed the merit of the Invention.

Mr Bell's motions were less rapid, for it was 1812 before his vessel was in motion upon the Clyde. He in his turn claimed the merit of the inventions and October 28th 1816 he published a most impudent and arrogant letter in the Newspaper (*Caledonian Mercury*) reprehending and reviling all former attempts—assuming the whole merit, even of instructing Fulton—and giving a gross misrepresentation of Mr Miller's experiements—even quoting Mr Miller himself as authority.[9]

The whole question of the development of steam navigation and the varying claims of the pioneers to credit and reward was to break out into increasingly bitter controversy in the 1820s. We have seen James Taylor, writing to William Huskisson, President of the Board of Trade, about Bell's "impudent and arrogant letter" and indeed Bell's proud claim that he had produced:

> . . . the first of steam-boats in Europe that answered the end, and is at this time upon the best and simplest method of any. . .[10]

certainly did not lack confidence although it hardly gave credit to Symington's pioneering, if eventually unsuccessful efforts.

Miller of Dalswinton's son, Patrick Miller Jnr also entered the fray. Whether he was motivated solely by filial piety or whether he also had an eye to a potential government premium he certainly attacked all around with some energy.

The unfortunate James Taylor is criticised for ever having induced his father to hire Symington:

> ... he [Miller Snr] used not unfrequently to reproach Mr Taylor rather smartly, for having ever thought of bringing such a person about him. This circumstance, as much as anything else, contributed to check the progress of steam-navigation in this country, from its introduction in 1788-9 till 1811, by damping my father's ardour. . .[11]

Miller Jnr goes on to suggest that his father had intended to build another steamship to journey from Leith to London. Taylor's account would not lend any support to this view and his account of Miller's growing interest in land improvements and consequent lack of resources to devote to steam navigation seems perfectly consistent.

Miller Jnr characterises Symington as no more than a hired helper:

> ... Symington, whom my father always considered in the same light as he did every other labourer or tradesman he employed about his different vessels. . .

and he is accused of making use of Miller's work in his 1801 patent, the inventions therein he says were:

> ... surreptitiously obtained many years after my father had made the discovery known to all the world, thereby rendering his letters-patent of no avail. . .

Quite how one "surreptitiously obtained" something that has been made "known to all the world" is not made clear. Miller Jnr's statement is also, of course, at variance with the statement published in the *Scots Magazine* after the Dalswinton experiment which pointed out that the engine used was "Mr Symington's new patent engine". Hardly, one might think, the language one would have used about one who was no more than a "labourer or tradesman".

However Patrick Miller Jnr's most vehement condemnation is reserved, perhaps understandably, for Henry Bell. Bell after all had actually brought the steamship into practical commercial service, an achievement which Miller Snr and Symington, for all their priority in development, had signally failed to accomplish, and might on these grounds be seen as the front-runner in the race for official recognition and reward. Miller wrote to the 2nd Viscount Melville, First Lord of the Admiralty, in April 1823 outlining his father's series of experiments and then noting that:

> A Mr Henry Bell of Helensburgh near Glasgow who I suppose had either been carpenter to Mr Fulton or his first steam worker and who probably knowing that my father is dead and was rendered before death long incapable of entering into any public controversy with him sets up a claim to the merit and has the modesty to ask a remuneration from the public for doing in 1811 exactly what I have just now shown your Lordship originated with my father in 1787. . .[12]

Patrick Miller Jnr continued his campaign with considerable vigour if, sadly, with some quite evident lack of balance. In a letter to the Edinburgh publisher Archibald Constable in 1825 he wrote of Symington that he:

> . . . had nothing to do with the matter more than merely as a mechanic applying the engine for my father had resolved upon the principle and announced publickly his intention of putting it to the test of experiment for months before he knew that such a person. . . Symington existed and most unfortunate it was for him for me and for others he ever had anything to do with him. . .[13]

Constable had sent Miller a copy of James Cleland's pamphlet on steam engines[14] and the reference therein to Bell provoked Miller to a remarkable outburst:

... the carpenter [Bell] whom he [Cleland] so highly eulogises knowing that my father is dead may suppose that his fame is a matter of indifference to his family and that therefore the opportunity may be favourable... to claim part of his reward for what he has no more pretension to than any swindler who has been committed in any our courts for endeavouring to approbate to himself the property of another by falsehood, fraud and imposition but I shall not peaceably submit to his being either decked in Laurel nor incased in Gold at the expence of one I feel called upon to defend from such abominable detractions...

Cleland commented, with a commendable degree of restraint, that he was:

... truly sorry at having been the unsuspecting cause of uneasiness to Mr Miller...[15]

and pointed out that he had no ulterior motive in backing Bell and that he had not written:

... for the purpose of bolstering up the Claims of a man, in whom I have no interest, nor of any Sinister Motive whatever.

But he insisted:

As to Mr Bell's Merits, as the first person in Europe who successfully impelled vessels by steam, or in other words, who first applied Steam Navigation to the purposes of commerce, I do not think there can be any doubt. And, of there being no essential improvement in the Machinery, since he placed his Boat the *Comet* on the Clyde in 1812, the Certificate of all the Engineers in this place bear ample testimony.

You are probably aware that as far back as 1816 I published an article, wherein I held out Mr Bell as the first person in Europe who had successfully impelled vessels by steam, and till now I never heard of Mr Miller's objections.

One interesting comment on Patrick Miller's merits and claim to fame as a naval architect comes from William Symington's letter to the *Edinburgh Courant* in 1825, written to counter Miller Jnr's earlier article. Symington points out that:

> ... no double or triple vessel, such as he recommended to the world, has ever been used up to the present date.[16]

Symington went on to observe that:

> ... his [Patrick Miller's] whole merit in making a steam-boat consisted in expending a certain sum of money, which enabled me to use a certain portion of my time, and to employ labourers under me to perfect that invention... He had the merit of laying out a sum of money necessary to make the first experiment, but he had not the merit of any invention.

This, it can safely be said, would be an argument likely to make Patrick Miller Jnr reach for his pen again!

Taylor was also active in the controversy and wrote once more, in August 1825, to Huskisson at the Board of Trade. Symington he quickly dismissed:

> Had he ever applied steam to the propelling of vessels previous to his being employed by Mr Miller, at my suggestion. . . ?[17]

Robert Fulton fared no better:

> ... his having seen Mr Symington's production and having had it fully and distinctly explained by Mr Symington himself a number of years before his own appeared in America totally destroys his claim as an original inventor.

While Bell's claims are dismissed angrily:

> ... and his pretensions appear to me to be the worst founded,

and the most impudent of all; for he not only had the advantage, in common with Mr Fulton, of seeing Mr Symington's vessel. . . but he had the additional advantage of knowing the successful result of Mr Fulton's attempt on the Hudson. . .

He concluded that:

> . . . the honour of the invention lies betwixt Mr Miller and me. . .

and argued that his position as a tutor in Miller's household and Miller's undoubted role as sole financier of the experiments meant that if there had been any doubt about the value of Taylor's contribution Miller could have with propriety claimed the sole honour. He pointed out that even though Mr Miller was:

> . . . quite an idolator of fame. . .

nonetheless Miller had never made any claim to exclusive credit for the invention. In fact, Taylor argued, Mr Miller:

> . . . was conscious that he was neither the sole nor the principal inventor of Steam Navigation; and had too much honour to appropriate what was not his own.

At the end of his long letter he points out, somewhat mournfully, that all these details should have been submitted long ago but that he hoped they were still in time to show that his claim was legitimate. If not then he could only:

> . . . lament that a combination of unfortunate circumstances, over which I possessed no control, has prevented me from reaping any benefit from a discovery, which in the hands of a more pushing, or more lucky individual, might have raised him to opulence and distinction.

Indeed opulence and distinction was to evade most of the principal players in the game. Bell certainly never achieved great financial rewards and William Symington died in London, according to his memorial, "in want", two years after Bell, in 1832. His plight was not helped by litigation in which he was involved. In December 1814 he opened proceedings to sue Henry Bell for breach of his 1801 patent on a "A new mode of constructing steam engines". Bell successfully defended the case, which went to the Court of Session, in March 1815.

The central feature of Symington's patent was that rotatory motion was produced without the interposition of a lever or beam but Bell's defence, drafted by advocate Robert Forsyth, pointed out that rotatory motion was, in fact, in Symington's design achieved in just this way.[18] This was not Symington's only journey through the courts. He planned, but did not apparently carry through, a joint action with Taylor against the *Clyde* steamboat owners and he was in his own turn sued in 1820 over a mine engine contract.

Taylor died just weeks after his last letter to Huskisson:

> . . . bowed down by infirmity and the fruit of a long life of disappointments. . .[19]

His widow received a government award of £50 in recognition of James Taylor's work.

Robert Fulton, who also spent much of his income on litigation to defend his patents and monopolies, and in consequence died in somewhat straitened circumstances, reaped the greatest long term, albeit posthumous, reward. The United States Congress in 1846 voted the sum of $76,300 for the relief of his heirs, an encouraging sum, even if provided 31 years after Fulton's death.

Robert Fulton, who started a commercial steamer service in
New York state in 1807
(Mitchell Library, Wotherspoon Collection)

*Helensburgh circa 1838, from an engraving by Joseph Swan
(author's collection)*

CHAPTER 5
Inn-Keeper & Provost

Down to the mid eighteenth century there had been, on the north shore of the Clyde, some eight miles downstream from Dumbarton a small fishing settlement called Milligs. The lands of Milligs had originally formed part of the estate of the McAulays of Ardencaple. The Chiefs of Clan McAulay had however, like many minor landed families, fallen on hard times and in 1700 their Milligs property was sold to Sir John Shaw of Greenock. Shaw's plans to develop the area came to nothing and he in his turn sold the estate.

The purchaser was a typical eighteenth century improving laird—Sir James Colquhoun of Luss. In 1776 Colquhoun decided that his best policy was to follow the example of many of his fellow landowners throughout Scotland and develop his new property of Milligs by erecting a planned industrial village on the site. The chosen method was to offer to feu, or lease, the lands. Specific encouragement was to be offered to "bonnet-makers, stocking, linen, and woollen weavers"[1] to settle in the new community. The new village was to be:

> . . . regularly laid out for houses and gardens, to be built according to a plan . . .

and had such amenities as a freestone quarry and:

> . . . for the accommodation of the feuars, the proprietor is to
> enclose a large field for grazing their milk cows, &. . .

This new planned settlement was called Helensburgh, in honour of Colquhoun's wife Helen Sutherland. The new town never became a notable centre of the textile industry and indeed had a somewhat slow and uncertain start to its development. As late as 1812, the year when the *Comet* first sailed, there were still only 61 feuars in the town.

Nonetheless by 1802 it was considered sufficiently well established to be granted burghal status to encourage industry and promote manufactures. Under its charter as a burgh of barony, that is a town granted its rights, constitution and status by the landowner or feudal superior (as opposed to a Royal burgh which was granted such privileges and powers directly from the Crown) Helensburgh was entitled to elect a Provost, Baillies and Councillors, hold markets and take such other measures as were the common practice of the day.

Despite such permissive powers no steps were taken to implement these rights until 12th September 1807 when, at a general meeting of the feuars "agreeable to their charter" Hendrey (sic) Bell was elected Provost, and Robert McHutcheon and Thomas Craig were elected Baillies and the council was completed by Daniel Colquhoun, Donald McFarlane, John Bramander, Charles Colquhoun, with John Gray and Robert Colquhoun as Town Clerk and Depute Clerk.[2]

The newly elected Council soon was involved in civic business. On 16th September 1807 the Council agreed to "make a road and side path through the city of Helensburgh" and 14th November 1807 saw it resolve that annual markets should be advertised and later that month Bell and his fellow councillors had a discussion on forming streets in the township. Ambitious plans were swiftly laid, and perhaps it is not too fanciful to think that one can detect in them the enthusiastic and scheming spirit of the first citizen. In November Bell and Baillie McHutcheon purchased ground from James Smith to build a town's house and in February of 1808 won the electorate's approval, by a majority of four votes, to proceed with a plan for the town's house, that is to say a municipal buildings or town hall, and markets.

However the progressive spirit would perhaps seem to have over-reached itself and despite receiving a tender from James Bramander to build the town's house for sum of £260, by 2nd May 1808 the Council had decided to give up the feu they had acquired for the town's house and authorised Provost Bell to make the best deal he could with the landowner, Sir James Colquhoun, on the surrender of the feu. The minutes go on to record the sad end of the Council's ambitions:

> I Hendrey Bell, Provost this day on behalf of the town of Helensburgh have hereby agreed with Sir James Colquhoun for the sum of eighty pounds and to give him up the feu purchased for the Town from James Smith.

However not all the plans of the first Council ended in quite such an embarassing fashion. At the same time as it backed out of the plan to build a civic headquarters it authorised the building of the market place for the new town.

There then followed a period of inactivity, or consolidation, or perhaps just bad record keeping. For whatever reason the Town Council Minute Book has no entries until 12th September 1808 when the second annual election meeting was held, resulting in the re-election of Henry Bell as Provost. Another quiet spell followed with little of note seeming to take place except the Council deciding on 3rd July 1809 to hold the burgh court on Tuesday 10th August at the new Theatre at 5pm and regularly thereafter on the first Tuesday of each month. Failure on the part of members of Council to attend court after due warning would lead to fines of 10/- in the case of the Provost, 5/- for baillies and 2/6 for absent councillors.

On 12th September 1809 the annual election saw the return of Bell as Provost, the only fixed point in an entirely new council.

John Gray continued to serve as Town Clerk but with a new Depute Clerk, Robert Bain. This may very well have been the same Robert Bain who was later to continue his association with Bell by becoming skipper of the *Comet*. The qualifications for ships' captains at this time were somewhat informal. It will be recollected that John Robertson had said that the *Comet*'s first skipper, William MacKenzie, had kept a school, so there is little inherent improbability in Robert Bain being able to act as Depute Town Clerk and later as master of the *Comet*.

From what we know from other sources about Bell's involvement and interest in water supply schemes it is reasonable to detect the influence of the Provost in the decision minuted on 16th May 1810 to provide the growing town with a supply of spring water. This, it was agreed, would first be led:

> into a proper reservoir;
>
> 2nd from thence conducted into two different branches the one down Sinclair Street and the other to be carried down James' Street and the same to be delivered from a pipe so as to admit of water stoups or barrel to be filled from the same;
>
> 3rd it is further agreed by said Magistrates that if the said proposal meet with the approbation of the majority of the Feuars that an assessment of one penny per pound of the valuation of yearly rent upon each feuar or proprietor in the village the first instalment to be paid the first of June curt. and the same assessment to be continued each year upon the first of June . . . until the said water be distributed through the town to the satisfaction of the Magistrates in time being.

This scheme was to be Henry Bell's last major project as Provost. At the annual election in 1811 he was not elected and he was to take no further part in civic affairs. Indeed he is not even listed as being present at election meetings of feuars. The Council, it is reported, had a policy of not re-electing a Provost or Baillie on more than two occasions without a break in office and the implementation of this policy would of course be sufficient to explain his demission from office.

However even the existence of such a policy would not explain why Bell ceased to take any part in civic affairs, at even the very modest level of attending election meetings. It seems likely that with the transfer of ownership of the Baths Inn to Archibald Newbigging and Bell's consequent change in status from a feuar, or landholder, in the Burgh to a tenant he may have ceased to qualify as an elector and as such ceased to be eligible to serve as a Councillor. It was not until 1828 that he again became a property owner in the town and thus would have regained a qualification as an elector.

Even if Bell had remained qualified one may speculate that he would not have been anxious to continue in office. His mind was now turning increasingly to his steamboat project, and he would surely have found difficulty in being able to spare the time for the minutiae of burgh administration. It is also very likely that his characteristically ambitious plans—the town's house, the water supply scheme, etc—were so far in advance of what the cautious and economical citizens of Helensburgh were prepared to support and pay for, that he might well have found himself being voted out of office had he been able to stand again.

However reluctant the burgesses of Helensburgh may have been to pay for their first piped water supply, the scheme that was entered into in 1810 was by no means as ambitious as Bell himself would probably have wished. Donald Macleod's *A nonogenarian's reminiscences of Garelochside and Helensburgh* were published in 1883 and, being written with the advantage of the author's uncle's personal memories of Bell, gives both an insight into Henry Bell's character and a typically colourful account of one of his many plans. In this case his scheme for a water supply for the town would have involved the creation of a reservoir in nearby Glen Fruin.

He had carefully surveyed the ground and pipe tracts, and taken levels and measurements. The plan involved the construction of a large reservoir on Kilbride Farm at a spot affording great natural facilities for such a work. One summer day, accompanied by a band of workmen, with chains, poles, flags, and spades, he appeared on the farm, and staked off the site of a proposed reservoir. The whole agricultural population of the glen turned out. They had a vague idea that Bell was uncanny, and curiosity led them to inspect the proceedings. McFarlane of Darling, the patriarch of the glen, demanded explanations, got them, and them remonstrated, on the ground that his meadow land, which was the best of his farm, would be put under water. "Precisely so". replied Bell, "but d'ye no ken that an acre of water's worth twa o' land. We'll droon the meadow, ay, an' I think ye'll see it far beyond that, for I intend it should rise nearly to the level o' your hearthstane." But there was a twinkle in his eye, indicating that he was quietly enjoying the fun of dismaying the honest farmer with dire forebodings of the safety of crop and household. From the proposed reservoir the pipes were to have been led by the Drumfork road, thence across Luss Road, and down the lands of Kirkmichael to the town.[3]

While Bell's steamboat activities would in themselves have provided one good reason why Henry Bell might not have been too reluctant to drop out of civic affairs in Helensburgh it must also be remembered that he and his wife Margaret had the Baths Inn to run. Although both participated in this, it would seem that Margaret carried the main burden of the management of the Inn—not, one would think, an easy task when married to a man like Bell who always seemed better at finding ways of spending money than ways of earning or keeping it. Donald MacLeod wrote:

> He was a child in the matter of money. Nothing delighted him more than to have his pockets full, not for its own sake, but solely that he might pay it away right and left. Well his men knew on a pay Saturday by the first glimpse of him whether he was in funds or not.

The reference to "pay Saturday" is a reminder that at this time workmen were normally paid fortnightly rather than weekly.

> If his purse was full he invariably whistled softly over the
> first two lines of "Logie o' Buchan", the one tune he was
> fairly master of. "There was a full head o' steam on," said
> one of his former men, "when the well-known stave was
> whistled; we were sure o' oor money." Bell's first desire was
> to get rid of it. It made him supremely happy to distribute it
> and this he did as quickly as possible. Often to the annoyance
> and sore inconvenience of his worthy wife, who managed
> the hotel and needed a share of it, was it discovered, when
> her claim was presented, that every penny was gone. And
> she had too much sense to scold, whatever she felt.[4]

The Bells' connection with the town seems to date from 1806
and he was receiving mail there in July and September of that
year. In July of that year the Register of Sasines records that Henry
Bell, Architect, of Glasgow, had feued on 29th May a piece of
ground lying on the south side of the road from Dumbarton to
the Kirk of Row.[5] This was the site of the Baths Inn. When
Margaret Bell died in April 1856 aged 86 her obituary notice
mentioned that she had been associated with the Baths Inn for
forty nine years. This would suggest a date of 1807 for the
opening of the Inn, which is certainly consistent with Bell
acquiring the land in 1806. Bell's last appearance in the Glasgow
Directories is for the issue of 1807, which again suggests that
their move away from the city took place in that year.

In an obituary of Margaret Bell, which, in some respects,
would seem to be rather unreliable the writer states that:

> In 1806 a Joint Stock Company projected the Baths Hotel at
> Helensburgh, and Henry Bell was somehow engaged to
> superintend the building. From some reason the projectors
> abandoned the speculation soon after the hotel was built,
> and it was purchased by Henry Bell.[6]

This remarkably vague statement is of course contradicted by the evidence of the Sasines and although there is nothing improbable in the story as presented there is no evidence to support it and the entry in the Register of Deeds recording Bell's disposition of the property to Newbigging speaks of:

> . . . the whole Baths, offices and Buildings erected by me on the said piece of ground . . .[7]

and would seem to settle the matter.

One assumes that if the Inn was built for Bell he would himself have acted as architect. Although established in business as an inn-keeper Bell did not give up his other enterprises and continued to work as an architect, as for example in the reconstruction of the Dalmonach printworks in 1812. Bell used a wide variety of descriptions—the re-registration document for the *Comet* describes him as "merchant of Helensburgh", while in a document for another ship in which he had an interest he is listed as "engineer in Helensburgh."

On September 6th 1809 Bell advertised in a London newspaper *The Star* that he had " . . . now nearly completed that part of his extensive plan of the Baths, which he intends at present to execute."

Certainly by 1810 the Baths Inn was well established. By May of that year, as we have seen, Bell had raised £2000 by disposing of the ground, Baths, offices and buildings to the Glasgow merchant Archibald Newbigging.[8] It seems reasonable to assume that Bell's motive in raising this sum by a mortgage on his property was to raise funds for his steamboat experiments and eventually the construction of what became the *Comet*. The terms of the agreement would suggest that there was little chance of redeeming the loan in the six months allowed, from Martinmas 1809 to Whitsun 1810. From that time forward the Bells no longer owned the Baths Inn though they continued to live there and manage it. The property, even in Bell's lifetime, passed through a number of owners and shortly after Bell's death one of his successors as Provost of Helensburgh, James Smith of Jordanhill,

Bell's business card
(Arctic Penguin, *Inveraray*)

rented it from its then owner, John Lang a Glasgow lawyer.[9]

In 1820 and 1821 the property was repeatedly advertised for sale—initially at £2000—the main building being described as a three storey block 50 feet by 51 feet. As shown in nineteenth century prints the Baths Inn was an impressive square whitewashed building. From the advertised description the Inn was a complex building of some substance and character—a large greenhouse or conservatory and a coffee room were features of the ground floor while the second floor apartments were 16 feet in height and included a 27 foot by 16 foot dining room. The baths department was supplied with numerous hot and cold baths. The west wing was sublet to a private lodger. The whole property was completed with stabling for 22 horses, a coach house for nine carriages, a shed for carts and gigs and the usual outbuildings.[10]

A further indication of the nature of the property is given by a *Greenock Advertiser* report of a fire at the Inn in 1821. In this report we are told that the fire broke out around 5pm on 12th June in the stables. The blaze destroyed all the outbuildings, but the horses in the stabling were saved although their harness was destroyed as were some pigs, doubtless kept to eat up the scraps from the hotel kitchens.[11] The semi-agricultural nature of the establishment is reflected in the inventory of Bell's personal estate submitted to the Sheriff Court at Dumbarton after his death— the first item on this included among the household furniture, clothes, and moveable effects, "farm stock & crops".[12]

The property clearly was not an attractive bargain at £2000 because in a later advertisement the upset price was reduced to £1800 [13] and at the end of 1821, clearly despairing of being able to sell the property, the Inn, described as:

Elegant commodious and well-frequented[14]

was advertised to let.

In 1821 John Galt, best known for his novels *Annals of the Parish* and *The Provost*, published in *Blackwood's Magazine* a new

104

episodic novel *The Steam-Boat*. This was published in book form in the following year and concerns the adventures and travels of Thomas Duffle and the tales that he is told by his fellow travellers. Mr Duffle's early travels are by steamboat around the Clyde and in one of his journeyings, which are depicted in a mock-heroic style, the intrepid Glaswegian shopkeeper reaches Helensburgh and we get an account of the Baths Inn and its health-seeking clientele. Galt's description is brief but valuable for its contemporaneity. It seems pretty certain that Galt, who for a time lived in Greenock, just across the Clyde from Helensburgh, was writing from first hand experience and knew both Bell and the Baths Inn.

> When I had ate my dinner and drunk my toddy at the pleasant hotel of Helensburgh, in which there are both hot and cold baths for invalid persons, and others afflicted with the rheumatics, and suchlike incomes, I went out again to take another walk. . .[15]

In the course of this walk Galt has his hero Duffle meet some Glasgow acquaintances, the McWaft's, and this passage has some interest as a contemporary reflection on the taste for seaside holidays which had been made possible by the coming of the steamboat and which would ensure the nineteenth century popularity of Helensburgh and other coastal resorts. The McWaft's had not settled at the Baths Inn but instead, like many other vistors, had taken a house in Helensburgh as their summer quarters. Their dwelling was a picturesque thatched cottage which however, was equally typically, damp and inconvenient. Duffle is invited to come back to the cottage to pay a visit and Mrs McWaft:

> . . . got out the wine and the glasses, and the loaf of bread that was blue-moulded from the damp of the house. . .

Duffle comments on the blue mould and suggests that the damp

105

which affected the bread could not be healthful to the ailing Mr McWaft, who replies:

> But the sea and country air . . . makes up for more than all such sorts of inconveniences.

Such enthusiasm for sea-bathing and the fresh air of the Clyde Coast was the making of the Baths Inn and ensured the growth and prosperity of Helensburgh during Bell's lifetime.

The problem of the water supply, both fresh and salt, for the curative baths so much sought after by the McWaft's and their contemporaries was solved readily enough, and it would seem, despite some statements to the contrary, without the use of a steam engine. The 1810 transfer of the property by Bell to Archibald Newbigging included the liberty to take:

> . . . a six inch bore of water from the Burn of Milligs below the termination of the mill race of the Milligs Mill and of conveying the water by a Ditch or Cast . . . down to the Baths for the use of the same or of driving a wheel to raise the Sea Water into the said Baths.[16]

The Milligs Mill was just over half a mile north west of the Baths Inn.

In 1823 Henry Bell placed an advertisement in the *Glasgow Courier* which suggests that the damage done in the 1821 fire had been made good and that the Bells were perhaps now looking more towards long term residents than to short-term holiday makers.

HELENSBURGH BATHS, INN AND HOTEL
THE HELENSBURGH BATHS, INN AND HOTEL have undergone a complete repair, and a good many improvements made to the Buildings, and are just now ready for the accommodation of the Public.
MR BELL can accommodate a few LADIES and GENTLE-

MEN as BOARDERS on reasonable terms. Also, TWO PRIVATE FURNISHED LODGINGS, TO LET by the Month—the West Wing of the Baths, and the Cottage; each of them consists of a Kitchen and five Fire Rooms and Closets.

H BELL is grateful for the kindness he has experienced, and the many favours he has received in his present situation, and hopes that the Public will countenance him as they think his merits deserve—as it will be his interest, so shall it be his study, to select his Wines, Liquors and every other article of the best quality. Comfortable Post Chaises, Steady Horses and Careful Drivers.

MRS BELL, himself and servants, are determined to pay unremitted attention to the conveniences and comforts of guests.[17]

The Bell family's other property interests in Helensburgh consisted of a house owned by Margaret Bell and her mother Janet Dykes [18] and six plots of land adjacent to the Baths Inn bought by Henry Bell in 1828. This latter transaction is of interest not least because of the description of the property:

Henry Bell, residing in Baths at Helensburgh, seized 14th June 1828 in six pieces of ground in East side of Bell Street and South side of Clyde Street. . .[19]

For once it would appear that the old adage about a prophet not being without honour, save in his own country, was disproved. Clearly Bell had, within his own lifetime, enjoyed the signal honour of having a street named after him. The street which commemorated him still remains and is now, more precisely, called Henry Bell Street and joins the main road through Helensburgh at the point where the Baths Inn, later to be renamed the Queen's Hotel, stood.

For most of its time under the Bells' control the Baths Inn ran smoothly and successfully, an achievement which almost certainly owed more to Margaret's talents than to Henry's. It swiftly established a position as the town's, and the district's, leading hotel. By the end of Henry Bell's life, and to no small extent due to his work, Helensburgh had grown to have a permanent population of around a thousand but boasted a summer population of three times that number.

A Gazetteer published in 1842, that is twelve years after Bell's death, but during Margaret Bell's long tenancy of the Baths Inn described the town in the following terms, which do seem to indicate a remarkable advance from Galt's description of twenty years earlier:

> As seen from the opposite shore, it is a town dressed in white, and seems to be keeping perpetual holiday; and, in certain and not infrequent combinations of shade and sunshine, it appears to be a miniature Venice, a city of the sea, resting its edifices, with their clearly defined outlines, on the bosom of the burnished or silvery waters. . . Most of the houses have been built solely or chiefly as sea-bathing quarters, and are not unworthy of their pretensions to be a pleasant summer-home to the families of the plodding and wealthy merchants of Glasgow. . . At the east end is an elegant and commodious edifice called the Baths, where the luxury of immersion in any degree of temperature may be enjoyed. The town . . . depends for subsistence almost wholly on its capacities as a watering-place; and while joyous, bustling, and full of life during the bathing-season, it fades away and languishes towards the approach of winter. . . The smallness and incommodiousness of the quay would seem to be a hindrance to its prosperity. Yet five or six steamers ply daily between it and Glasgow during 7 months of the year, making each three trips, one up and two down, or two up and one down in the day; and even during winter, 2 or 3 make daily trips, and keep up the communication; while there is almost hourly communication with the opposite port of Greenock, and thence by railway to Glasgow.[20]

However, during the latter part of Margaret Bell's long life the business of the Baths Inn declined and some glimpses of the operation of the Inn tend to confirm this. At the time of the 1851

Census Margaret was still running the hotel with the assistance of 9 live-in staff—a work force ranging from a cook, a dairy maid and a waiter to an errand boy. We may presume that there would have been additional non-residential staff, cleaners for example. However on the Census day the hotel could only boast of four guests, specifically two visitors and their two servants, which was hardly full occupancy for the eleven guest bedrooms— admittedly the Census was not taken at the height of the summer season. Although this represented a decline from the Inn's much busier earlier days, nonetheless the visitors that came to the Inn were of a very select nature.

In September 1853 the local newspaper recorded among the new arrivals Capt Gwynn of the 2nd Life Guards and Mrs Gwynn as well as the Lord Advocate, Scotland's chief law officer.[21]

The writer of Mrs Bell's obituary in the local newspaper commented on the popularity of the Baths Inn, recalling the period immediately after the commencement of the *Comet* service in 1812:

> From that time till her death Mrs Bell was hostess of the Baths Hotel, an inn of almost world-wide celebrity. It was for many years almost the only inn on the Clyde at all deserving the name; and not only was it a favourite resort for Glasgow people, who came down to the coast for a week or two, but for crowds of strangers, from all parts of the world, interested in Henry Bell and his proceeedings. Helensburgh did not then include more than four or five indifferent houses, and the Baths, consequently was almost the only place of accomodation for strangers and visitors.[22]

The reference to Helensburgh's "four or five indifferent houses" usefully confirms the accuracy of Galt's 1820s reference to the holiday-making McWaft's seaside residence as a "thackit house . . .damp and vastly inconvenient".

Mrs Bell was, as the obituary writer suggests, a very popular hostess, and long after Henry's death in 1830 the Baths Inn was still a place of pilgrimage for those interested in the story of the

Comet and Bell. Indeed, because of the relics of the great man which his widow preserved, it became something of a shrine to his memory.

One distinguished traveller who made the obligatory visit to the Baths Inn was Harriet Beecher Stowe, the author of *Uncle Tom's Cabin*, who describes in her autobiography a visit in April 1853:

> At the village of Helensburgh we stopped a little while to call upon Mrs Bell, the wife of Mr Bell, the inventor of the steamboat. His invention in this country was at about the same time as that of Fulton in America. Mrs Bell came to the carriage to speak to us. She is a venerable woman, far advanced in years.[23]

The local newspaper recorded this visit and noted that the American writer was:

> . . .presented with two engravings of the *Comet*, the first European steamer, by Mrs Bell, the venerable hostess of the Baths, towards whom she evinced particular attachment.[24]

Mrs Bell's returns to the Censuses of 1841 and 1851 offer both an interesting reflection on the reliability of the Census as a historical record and on the age-old custom of women being coy about their age. In 1841 Mrs Bell reported herself as being 58 years of age. Ten years later she had, remarkably enough, become 72 years old. In reality, if her death certificate and the obituaries can be relied upon, in 1841 she was no less than 71 years old. In other words at the time of the 1841 Census she claimed to be thirteen years younger than she really was, while ten years later she only felt able to claim nine years grace.

Margaret Bell, born near Lesmahagow, was described in her obituary notice in the *Dumbarton Herald* as being:

> In her girlhood . . .the belle and toast of that rural parish,

110

and till the very verge of her long life the evidences of her
early beauty and vivacity were strongly marked.[25]

From this description one may imagine that her deception of the
Census enumerator over her age was quite credible.

Margaret died on 30th April 1856 at the great age of 85, her
death certificate giving the cause of death as "old age and general
debility". The *Glasgow Herald* observed that she was:

> ... the widow of the celebrated Henry Bell, the man who
> amid difficulties, derision and discouragement was the first
> to propel a vessel by steam through British waters... She is
> to be interred on Wednesday first at one o'clock pm and we
> need scarcely hint to the steamboat owners on the Clyde that
> on that day they should carry their flags half mast high as a
> mark of respect to the memory of the deceased lady and her
> celebrated husband.[26]

The Clyde Trustees grant of £100 per annum to Henry was
continued to Margaret after his death, a decision which must
have provided a welcome regular income to Mrs Bell. It would
also seem reasonable to assume from her life-long continuation
in the tenancy of the Baths, despite its obvious decline in custom,
that its owner, James Smith of Jordanhill, was by so doing,
ensuring the financial security of the widow of Helensburgh's
most famous figure. The local newspaper commented that:

> She never gave occasion to those unfounded rumours which
> got circulated regarding neglect and poverty in her old age.
> Mrs Bell never complained of herself, though she often did
> of the way in which her husband had been treated and his
> memory disregarded. She treasured up with great care every
> memorial of the labours of her husband; and we believe a
> great deal of interesting correspondence connected with the
> early progress of steam navigation, which she would not
> allow to be published during her life, will now be given to
> the world.[27]

Sadly for the biographer this "great deal of interesting correspondence" would seem to have disappeared and often the only indication of what there might have been comes at second hand through Edward Morris's *The Life of Henry Bell.*

CHAPTER 6
"...Mr Bell's Scheme..."

On 2nd April 1813, with Britain still embroiled in a twenty year long war with France, Henry Bell wrote to Lord Melville, First Lord of the Admiralty, enclosing his newly published pamphlet *Observations on the Utility of Applying Steam Engines to Vessels...* In what is undeniably a somewhat mysterious and confusing, though typically self-assertive, letter Bell offers to make available to the Admiralty:

> ...a Machine that one of these small portable Engines with two men could keep off any ship's crew from boarding any Vessel with any boarding implements whatever, and even at the distance of one hundred feet, none of them could stand on their own vessel's deck, and the nearer, the effect of the machine will be the greater. It takes no lives but acts in one moment.
>
> Second I could construct a Vessel Bomb-proof to be wrought by steam to go into any harbour and act even in the wind's eye at the rate of six to seven miles an hour, and one of these machines as above would keep any from boarding her, or she could cut out, or burn or otherwise destroy Vessels in harbour and when any way entangled she could come out at pleasure. You will find on a trial that one or two of these Vessels with a fleet when blockading any harbour or river will be of great service, as they, when the rest of the

Vessels are becalmed could go in and do great execution, or could keep all vessels from running along the coast as they draw so little water.[1]

There would appear to be three quite separate ideas contained in this letter. The most obvious of these is the proven fact of steam propulsion—"a vessel . . . to be wrought by steam"; secondly, the "Bomb-proof" vessel and finally, and most intriguingly, the machine utilising a portable steam engine that would render British ships unboardable and would render the enemy crew incapable of standing on their own decks even at a distance of 100 feet—"It takes no lives but acts in one moment."

Quite what this third device was is unclear. Perhaps the most likely explanation, in view of its warranted non-lethal nature, is a high pressure hose powered by a portable steam-engine. That such devices were considered by other contemporaries is demonstrated by an article written by Captain William Bain on steam navigation. In this he suggests the military advantage of steamships:

. . . squirting hot water. . .[2]

at the enemy.

It would seem that the long-suffering members of the Admiralty Board, who had been favoured with a steady supply of secret weapons, grand strategies and a variety of hare-brained ideas all guaranteed to win the war, were unimpressed or perhaps were simply unclear as to what was on offer. Bell's letter is endorsed, apparently by Lord Melville himself:

This is not by many the first proposal of a Vessel to be driven by steam, and in this as in all other inventions, the request is that John Bull should pay the piper. Steam Vessels may make progress in Canals and still water but to contend with the Atlantic Ocean and other seas the projector may as well attempt to "bottle off" these great national waters. . .

114

His Lordship makes no reference to the merits of the anti-boarding machine or the allegedly bomb-proof vessel.

The Admiralty's innate conservatism and deep suspicion of steam-ships was to continue for many years and was indeed natural enough in the rulers of what was the world's most powerful navy. The status quo suited the Royal Navy for the good reason that new ideas like steam, submarines and torpedoes were more likely to favour the weaker side, disrupt the balance of power and thus upset British maritime supremacy. However the comment that steamships might do well enough on canals but not in deep waters is, even for this source and this time, remarkably short sighted. Bell's letter had continued:

> If your Lordship would chuse to order me I would come up to London with my Steam Vessel called the *Comet*, and your Lordship and the Right Honourable Board could see her.

The *Comet* had been sailing regularly on the Clyde since August 1812 and she had been joined on the river and firth by the *Elizabeth* and the *Clyde* steamers. If Bell had been asked to fulfill his offer and produce the *Comet* before the Admiralty Board and had succeeded in sailing her from the Clyde to the Thames then the Lords of the Admiralty would surely have had a satisfactory demonstration and might well have been obliged to concede the practicality of sea-going steam. Whether the small and under-powered *Comet* could have made the passage is of course another question but one might think that the Admiralty could have shown sufficient intellectual curiosity and concern for the nation's maritime power to put Bell's offer to the test.

Lord Melville's cynical comment:

> . . . in this as in all other inventions, the request is that John Bull should pay the piper. . .

perhaps was as influential a reason as any other for ignoring Bell and was doubtless provoked by Bell's concluding sentences:

115

> I have laid out upon the different improvements upwards
> of Two thousand pounds sterling and I hope your Lordship
> and the Right Honourable Board of Admiralty would give
> me great credit when you saw her. I hope your Lordship
> will let me know as soon as you find it convenient.

In the event, the Board, in Melville's words, decided not to "trouble him" and the *Comet* never made the passage to Westminster.

Bell's application to the Admiralty in 1813 was at least based on an unassailable position as the proprietor, designer and promoter of a practical commercial steamship. He is however supposed to have made two much earlier approaches to the Admiralty on the subject of steam navigation, in 1800 and 1803—both apparently falling on equally stony ground. In Bell's autobiographical sketch written, for Edward Morris's benefit, in October 1826 he tells the story of these earlier rebuffs by the Admiralty.

> In 1800, I applied to Lord Melville, on purpose to show his
> Lordship and other members of the Admiralty, the
> practicability, and great utility, of applying steam to the
> propelling of vessels. . . My whole scheme was laid before
> the Admiralty for their deliberation. After duly thinking over
> the plan, the Lords of that great establishment were of
> opinion that the plan proposed would be of no value in
> promoting transmarine navigation! In 1803 I made a second
> application to the same high quarter, and from Lord Melville
> received a reply that nothing would be done there. They had
> no faith in steam navigation. The late Lord Nelson thought
> otherwise: he saw at once its mighty power. He rose and
> emphatically said, 'My Lords and gentlemen, if you do not
> adopt Mr Bell's scheme, other nations will, and in the end
> vex every vein of this empire. It will succeed,' added the
> gallant Admiral, 'and you should encourage Mr. Bell.[3]

This version of events, including the perceptive support of

Nelson, appears in the *Dictionary of National Biography* and Robert Chambers' *Biographical Dictionary of Eminent Scotsmen*. It is a splendid story with an all-star cast: Melville, the short-sighted First Lord; Henry Bell, the unappreciated genius; and Horatio Nelson, the national hero and the only man present with the vision to see the benefits of the Scottish engineer's proposals.

Sadly there is not a scrap of evidence to suggest that this story is anything more than the product of Henry Bell's ever-fertile imagination.

The Admiralty Board Minutes[4] for 1800 exist in an indexed fair copy but have no reference to any discussions or deliberations on the subject of steam navigation. The 1803 Minutes[5] only exist in rough form and consist of many hundreds of hurriedly written slips of paper bound up in two volumes. Without a great deal of very careful scrutiny it would be impossible to state with absolute confidence that the Board never discussed Mr Bell's views on steam navigation. However the efficiency of the Admiralty Secretariat fortunately has provided other independent means of checking on these supposed applications of 1800 and 1803. All letters coming into the Admiralty were recorded and indexed by subject and correspondent. Only one letter from Henry Bell is recorded between 1800 and 1815.[6] This was the letter of 2nd April 1813 quoted above. In addition to being indexed by the Secretary's department, such miscellaneous in-letters, or as the Admiralty categorised them "Promiscuous Letters", were saved and filed by the initial letter of the correspondent's surname. The 1800 file was apparently destroyed in 1808 but the 1803 file is still extant[7] and contains no Bell correspondence.

One can therefore say with some confidence that, as far as can be traced from the official records, the purported submissions by Bell in 1800 and 1803 did not take place. The probability of such a statement being correct is reinforced by the survival in the Promiscuous Letters file for 1813[8] of Bell's letter of 2nd April and the accompanying pamphlet and the appropriate annotation in the Secretary's Index for that year[9], thus suggesting that the recording, indexing and preservation of correspondence was being carried out efficiently. One may thus feel that the non-existence of any trace of the 1800 and 1803 applications can not be attributed simply to poor record keeping. In short it seems highly improbable that the preserved correspondence, the Board Minutes and the Secretary's Index would all fail to show some trace of these alleged submissions.

There are however a number of other indications that Bell's account is simply fictitious. The Lord Melville to whom he wrote in 1813 and who was so dismissive of his offer was Robert Saunders Dundas, 2nd Viscount Melville (1771-1851), the son of Henry Dundas, 1st Viscount Melville (1742-1811). Robert Saunders Dundas served at the Admiralty as First Lord from 1812 to 1827 and again between 1828 and 1830. The 1st Viscount Melville, whom Bell one assumes refers to in his account quoted above, was also First Lord of the Admiralty—but only from 1804 to 1805. While Lord Melville had admittedly earlier served as Treasurer of the Navy if he had dealt with Bell in this capacity over the purported 1800 application it would have been as plain Henry Dundas and not as Lord Melville, his elevation to the peerage only coming in December 1802. While the wrong name might be put down to confused memory a more telling argument against the truth of Bell's story, which has Melville figuring in both the 1800 and 1803 applications, is the fact that in 1803 Dundas/Melville was not in the government in any capacity. The First Lord of that period was the Earl Spencer. In any case, although a letter to Bell from the Board of Admiralty might have been sent in the name of the First Lord, the correspondence would have normally been signed by the Secretary to the Board.

The claimed support of Admiral Lord Nelson for the second application is also extremely dubious. In the early part of 1803 Horatio Nelson was indeed in England. This was the period of the Peace of Amiens and Nelson was unemployed and on half pay. When war broke out again in May 1803 he was appointed as Commander in Chief in the Mediterranean and left on 20th May to take up this command. He did not return to England that year. If Bell's supposed letter was discussed at the Admiralty with Nelson present then it must obviously have been before May 20th. However Nelson never at any time formed part of the Board of Admiralty. Why therefore he, albeit a national hero but technically just one of the scores of half-pay flag-officers on the Navy List, should have been invited to join in the Board's alleged deliberations on this matter is not explained. Nelson's participation in the Admiralty Board would of course have been highly irregular.

There is of course a further internal inconsistency in Bell's account. He states in the context of the 1803 application that he received a reply from Lord Melville stating that nothing would be done there. Setting aside the fact that Melville was not at the Admiralty at this time (one could perhaps with charity imagine that over the span of 23 years Bell had managed to confuse his First Lords) the implication is of a letter from the Admiralty rejecting his offer. However Bell goes on to give a colourful and seemingly verbatim account of Horatio Nelson's views. One can hardly imagine that any First Lord or Secretary to the Admiralty in writing an official letter of rejection would compromise the official position by gratuitously giving a detailed report of the dissent of a person who was not even a member of the Board. Bell makes no suggestion that he was present at the Admiralty when these alleged deliberations took place and indeed the whole tenor of his account indicates a written response:

> and from Lord Melville received a reply that nothing would be done there . . .

but still he quotes Nelson's remarks as if he had himself heard them.

The best that one can say for this story is that it is just possible that, if this improbable incident had actually taken place, there might have been a leak from the Admiralty and a garbled version of it found its way back to Bell. However all the weight of evidence makes it impossible to avoid the conclusion that this whole matter of the 1800 and 1803 submissions is entirely Bell's invention. It is surely highly significant that in his letter of 1813 he makes no reference to any previous approaches. One is entitled to think that, writing in 1813 from the security of his position as the proprietor of a practical steamship, he would have been likely to mention his earlier suggestions. Indeed from all that we know of his character it seems reasonable to suggest that Bell would have been unable to resist taking the opportunity of pointing out to the Admiralty that he had proposed such a development years before and that his prescience had been vindicated.

However no such reference to an earlier proposal is made in his letter and the story of Bell's Admiralty application or applications cannot be traced to an earlier source than Bell's letter on steamboats printed in the *Caledonian Mercury* in 1816. This version, while somewhat vague in detail, bears only very slight resemblance to the more colourful story quoted from above and written ten years later for his biographer's benefit. The 1816 version goes:

> . . . a few years ago I applied to my own Government, for the purpose of recommending steam-boats being made useful to our Government in time of war. After I had spent a great part of my time in making models of what I conceived would be of great service to us as a nation, for protection, and of great advantage for transporting troops from one part of the island to another: and the said troops, with their arms and implements of war, might with ease be carried from Leith to London in three days, and at the moment of their arrival would be quite fresh for action, if required—this, and many other improvements, I gave the Board of Admiralty, and even was flattered by the said Board for some time; but at length I got a letter from their Secretary, that it would be no use to them, after I had spent a great deal of both cash and time. This was all I received as the reward of my exertions; and when denied by my own Government, I sent out to America, to my old friend Mr Fulton, a full description of what I wished my own Government to embrace, and you have all gotten account of the steam frigate, called Fulton the First.[10]

This account would clearly seem to relate to the 1813 letter. It is interesting to note that Bell says that he received a reply from the Secretary to the Board, surely a more likely procedure than the suggested communication from the First Lord which he was to claim for the 1800 and 1803 applications.

Bell's acquaintance with Robert Fulton (1765-1815) most probably dates from Fulton's stay in Britain after 1804 rather than from the American's earlier stay commencing in 1787, when he came to London initially to pursue his career as a painter of miniatures. In 1800 Fulton was in France planning the construction of the *Nautilus*—designed to sink British warships— hardly a propitious moment for the patriotic Bell to offer his advice and counsel. The reference to Fulton as "my old friend" and the mention of the frigate *Fulton*, which was launched in 1814, would seem to reinforce the identification of this account as a narrative of Bell's 1813 application to the Admiralty. This view is strengthened by Bell's dating of the incident as taking place "a few years ago". The letter's date of 1816 would make "a few years ago" a reasonable description of 1813, but is hardly a credible way to describe 1800 or 1803.

The account Henry Bell gives of his contribution to the development of steam navigation in this *Caledonian Mercury* letter is chronological and he refers to earlier correspondence with Fulton, then to the building of the *Comet* and only after this to his offer to the Board of Admiralty. If then it can be accepted that this account relates to Bell's April 1813 letter it is very significant that nowhere in this lengthy description of his services to the progress of steam navigation does he make any claim for more than one submission to the Admiralty or to any earlier submission than that described. At no point in Bell's career was undue modesty a discernible feature of his character. It is hardly conceivable that he could, in giving to the public a statement outlining his role in the rise of steam navigation, have failed to mention these two earlier, and if true, remarkably far-sighted submissions.

It is also worth noting that in the pamphlet he wrote in

1813 [11], which gives an account of the development of steam navigation, and which accompanied his April 1813 letter to the Admiralty, there is no reference to his making proposals to the Admiralty on this subject in the years 1800 or 1803.

We may accept then that Bell only made one application to the Admiralty—the submission made in 1813 and described in his *Mercury* letter of 1816 more or less factually. (Although it is hard to square the dismissive written comments of Melville about John Bull paying the piper with Bell's statement that he "was flattered by the said Board for some time.") Why, ten years later, did he apparently invent the 1800 and 1803 applications? It should also be noted that this story of persistent attempts to move the Admiralty was echoed in his 1829 Memorial *To the Noblemen, Gentlemen and Freeholders of the County of . . .*[12] This document also repeats the tale, told to Morris, of Bell circulating copies of his 1803 proposals to ". . . all the Emperors and Crowned Heads in Europe, and also to America."

His motive for lying can surely only have been to gain additional credit by placing his involvement in the development of steam navigation at as early a date as possible. His autobiographical letter to Morris was written late in life, in 1826, at a time when he was trying to obtain a government reward for his work and when other parties—such as Patrick Miller Jnr and James Taylor—were vigorously asserting the primacy of the Dalswinton steamboat, attacking what they saw as Bell's pretensions and staking their claims to any credit or award that might be going. Equally the 1829 Memorial was an attempt by the financially embarrassed Bell to win the support of the gentry of the Scottish maritime counties for a government grant for his work.

The confusion over the First Lords could be simple forgetfulness but is much more likely to be inept lying. Throwing in Horatio Nelson as a supporter of the cause looks very much like over-egging the pudding. Both were of course fairly safe lies. The 1st Viscount Melville and Nelson were both dead and Edward Morris, a sympathetic friend and supporter, was hardly likely to check Bell's account against the Admiralty archives.

The exact nature of Henry Bell's relationship with Robert Fulton is, like so many aspects of the Scots engineer's life shrouded in mystery and controversy. One contemporary account[13] by William Bain, Commander of the *City of Edinburgh* steam packet, even claims that Bell crossed the Atlantic to offer his services there and then returned to Scotland to build the *Comet*. There seems to be no other evidence for this and Bell himself never seems to have published such a claim and other contemporary writers, such as James Cleland, refuted it. Again it is hardly in keeping with what we know of Henry Bell's character to believe that he would have been reticent about such a great adventure or that he would have failed to tell his biographer and the world of such a journey. William Bain's position as an authoritative source of information on Bell's movements is not clear—although it is of course possible that he may have been a relation of Robert Bain, the long-serving skipper of the *Comet*.

Robert Fulton, who was born in Pennsylvania in 1765, came first to Britain in 1787. While in Britain he became known to Charles Stanhope, the 3rd Earl Stanhope (1753 – 1816), an enthusiastic scientific and engineering experimenter, and to Francis Egerton, the 3rd Duke of Bridgewater (1736 – 1803), the great canal promoter. His original purpose in travelling to Britain was to study and practice art but he quickly became engaged in a variety of engineering projects, including such diverse work as inclined planes for canal navigation, flax spinning and marble sawing. As early as 1794, while residing in Manchester, he inquired of Boulton and Watt the price of a 3 or 4 hp engine to be fitted in a boat[14]. In 1796 he published his *Treatise on the improvement of Canal Navigation.*

In June 1797, with Britain at war with Revolutionary France, Fulton crossed the Channel and stayed there for some seven years. One of his projects during this time was the construction of the *Nautilus* submarine. This, in the absence of French official support, he built at his own expense and demonstrated to French naval, scientific and governmental circles on the Seine in 1800. Later *Nautilus* was taken to the Channel and carried out a number of abortive attacks on British naval vessels blockading the French coast.

During his French period Fulton met Robert Livingston, the United States ambassador to France. Livingston had earlier secured for himself a government monopoly of steamboat navigation in New York state and the two men went into partnership to exploit this monopoly. Fulton, during his stay in France, conducted trials with a model steamboat and then in 1803 built a full scale steamboat which ran trials on the Seine in August of that year, a demonstration which the offical government newspaper the *Journal des Debats* described as "un succès complet et brillant".

Despite this triumph he failed to secure the interest of the French authorities and Fulton returned to Britain in 1804. With what could be described as either a remarkably even-handed enthusiasm for technological development or a complete lack of political scruple he promptly entered into discussions with the Admiralty on submarines and torpedoes. These discussions resulted in a secret agreement [15] being drawn up, in July 1804, between Fulton and HM Government in the persons of William Pitt, the Prime Minister and Melville, First Lord of the Admiralty. Fulton's scheme for destroying the French invasion fleets lying in the Channel ports would, if it had been successful, have earned him a minimum of £40,000. Attempts were made to use Fulton's torpedoes to attack shipping in French harbours, but with as little success as the attacks by the *Nautilus* on British shipping.

The parties to this contract each appointed commissioners to examine the plans—the government commissioners included John Rennie—which provides a possible, albeit indirect link with Bell. Fulton's plans were not well received by most of the naval authorities, and the defeat of the combined French and Spanish fleets by Nelson at Trafalgar in October 1805 removed the immediate threat of French invasion and Fulton's plans perforce reverted to the peaceful exploitation of steam power.

Samuel Smiles in his life of John Rennie gives a colourful, if unsympathetic, account of the trials of Fulton's plans:

> When Fulton proposed the scheme of his famous torpedo for blowing up ships at sea . . . Earl Stanhope made so much noise about it in the House of Lords, that a commission, of which Mr Rennie was a member, was appointed to investigate its merits . . . it was determined to afford him an opportunity of exhibiting the powers of his infernal machine, and an old Danish brig, riding in the Walmer roads was placed at his disposal. He succeeded, after an unresisted attack of two days . . . in blowing up the wretched carcass, and with it his own pretensions as an inventor.[16]

While back in Britain between 1804 and 1806 Fulton would seem to have visited Scotland, met William Symington and inspected the rejected *Charlotte Dundas* lying in a canal basin on the Forth and Clyde Canal. Symington's own version of this meeting places the date of this contact in 1801, but as Fulton was in France at this time it would seem clear that Symington's recollection (he was admittedly writing some twenty-five years after the event) was at fault as regards the date.

> In July 1801 . . . a stranger came to the banks of the canal and requested to see me. He very politely announced himself as Mr Fulton, a native of North America, and told me that he intended to return to his native country in a few months, but having heard of the steamboat experiment he could not think of leaving the country without waiting upon me in the hope of seeing the boat and machinery and procuring some information as to the principles upon which it was moved.[17]

Symington, it would seem, supplied Fulton with all the information he required, fired up the *Charlotte Dundas* and took the American engineer for a cruise on the Canal. It is possible that Bell and Fulton first met at this time. James Taylor certainly states that they visited the *Charlotte Dundas* jointly,[18] although it seems from Bell's statement that they had corresponded earlier. What does not seem to be clear is quite how Fulton originally was put in touch with Bell.

Bell's letter in the *Caledonian Mercury* gives a version of his first dealings with Fulton.

> I observed in your paper lately a paragraph respecting steam boats, in which the Americans claimed the right to the said discovery, which is become of so much utility to the public. On this account I propose to give you a full statement of what I conceive to be the truth, and will venture to state to the public an account of what I know of it . . . to give you a more correct account of the manner Mr Fulton, the American engineer, came to the knowledge of steam-boats, that gentleman had occasion to write to me about the plans of some machinery in this country, and begged the favour of me to call on Mr Miller of Dalswinton, and see how he had suceeded in his steam boat plan; and if it answered the end, I was to send him a full drawing and description of it, along with my other machinery: this led me to have a conversation with the late Mr Miller and he gave me every information I could wish for at the time; and I told him where, in my opinion, he had erred, or was misled by his engineer, and even, at the same time, I told him that I intended to give Mr Fulton my opinion on steam boats; and the friends of Mr Miller must have amongst their papers Mr Fulton's letter to me, for I left it with Mr Miller. Two years thereafter I had a letter from Mr Fulton, letting me know he had constructed a steam boat from the different drawings of machinery I had sent him out, which was likely to answer the end, but required some improvement on it—This letter I sent to Mr Miller, for his information, which must also be among his papers. This letter led me to think of the absurdity of writing my opinion to other countries, and not putting it into practice myself in my own country and, from these considerations, I was roused to to set on foot a steam-boat, for which I made a number of different models, before I was satisfied in my own mind. When I was convinced that they would answer

126

the end, I contracted with Mr John Wood & Co. ship-builders in Port Glasgow to build me a steam vessel. . .[19]

Edward Morris's version of the relationship between Fulton and Bell doubtless owes much to his conversations and correspondence with Bell.

About the year 1806, Mr Fulton was in England and Scotland. He saw the boats which Symington, Taylor and Miller, had built, which, however valuable as experiments, were of no practical utility. They could not venture on rivers, much less brave the roaring sea. Mr Fulton was much with Bell, and being a talented, well educated, shrewd man, he picked up much information on this great system; and Bell's models were those he adopted when he re-crossed the Atlantic; but I am sorry to be compelled to add, he forgot the Scotch engineer of Helensburgh, and took all the merit to himself.[20]

In 1824, Bell gave yet another version of the story in a letter to Morris.

I this morning was favoured with your letter, and reply to your inquiry respecting the late Robert Fulton, the American engineer. His father was a native of Ayrshire; but of what town, or district there, I cannot say. He went to America, where his son, Robert, was born, who was well educated, and showed an early inclination for engineering. He came at diferent times to this country, and stopped with me for some time. He published soon afterwards a Treatise on Canal Declining Railways . . . Mr Fulton published this work in England, in 1804; and on his way to France, called on me; and also when he returned. He was employed by the American government to come to England, to take drawings of our cotton and other machinery, which quickened his desires after all the engineering branches; these he took up very quickly. . . When I wrote to the American government on the great importance of steam-navigation, and its

127

admirable adaptation to those noble rivers, they appointed
Mr Fulton to correspond with me. From this matter-of-fact
statement, you see the Americans got their first insight of
this system from your humble servant.[21]

Bell's reference to writing:

> . . . to the American government on the great importance of
> steam-navigation. . .

is presumably a reference to his claim to have circulated copies
of his thoughts on steam navigation to governments around the
world in 1803.

One obvious point to draw from these rather confused
accounts is that, as we know Fulton had enquired, in 1794, from
Boulton and Watt their price for a steam engine to be fitted to a
boat, Bell can hardly have, as he suggests, taught the American
all about the subject. There is no reason to doubt that Fulton
had, through his contacts with Lord Stanhope and later in the
course of his work in France, given much thought to the question
of steam navigation. Equally there is no reason to doubt that
Fulton, who had something of a reputation as a plagiarist, took
every opportunity to pick anyone's brains who had something
to offer him.

John Rennie had a very low opinion of Fulton and writing
to Sir John Barrow (1764 – 1848), Secretary of the Admiralty, in
1817 and enclosing a copy of Fulton's book, he said:

> I send you Mr Fulton's book on Canals, published in 1796,
> when he was in England, and previous to his application of
> the steam engine to the working of wheels in boats. On the
> designs (as to bridges, etc) contained in that book, his fame,
> I believe, principally rests; although he acknowledges that
> Earl Stanhope had previously proposed similar plans, and
> that Mr Reynolds of Coalbrookdale in Shropshire, had
> actually carried them into execution; so that all the merit he
> has —if merit it may be called—is a proposal for extending

the principle previously applied in this country. . . I consider Fulton, with whom I was personally acquainted, a man of very slender abilities, though possessing much self-confidence and consummate impudence.[22]

This comment on Fulton does serve to underline the point that there was a significant amount of interplay of ideas and that ideas and inventions were not being developed in watertight compartments—Bell, Miller, Fulton, Symington, Stanhope, etc Stanhope had consulted John Rennie on steam navigation at a time when Bell was working with Rennie in London. Bell talked with John Robertson about steam power for four years before building the *Comet*. Robertson and Bell saw the *Charlotte Dundas* and presumably had, and took, the same opportunity to speak to Symington about her as Fulton did.

Harvey and Downs-Rose in their life of William Symington[23] point out that there is no evidence in the biographies of Fulton to suggest that he ever consulted Bell but suggest that Bell possibly wrote to Fulton seeking advice, and reinforce this opinion by quoting Patrick Miller Jnr's views. Patrick Miller wrote to the 2nd Viscount Melville in 1823 relating how his father Patrick Miller Snr had commenced a series of experiments in steam navigation:

In 1787 he published a short treatise on his double and triple bottomed vessels. . . He intimated his intention of beginning his experiments that summer by the application of the steam engine to them. About this time he transmitted a copy of his book to all the sovereigns of Europe by their Ambassadors at our Court except to Louis 16 to whom I had the honour of personally presenting it as I was then residing at Paris and a copy was sent to General Washington the President of the United States, with a few more to some of the leading individuals there as he conceived his plans of naval architecture peculiarly well suited to their waters.

It was the hint thus thrown out which Mr Fulton got sight of on its arrival in America that led him to make strait enquiry concerning the original of my father's experiments and which as I am informed brought him to Britain to obtain from him the most accurate instruction in the principles of this most important subject after which he returned home taking with him his first engine by his advice from Messrs Watt & Boulton and in 1806 he put the plan into execution on the Hudson River. Having now briefly traced to your Lordship the modern history of steam navigation at least as far as my Father is concerned I come to the point on which I would wish to beg the honour of your support.

A Mr Henry Bell of Helensburgh near Glasgow who I suppose had either been carpenter to Mr Fulton or his first steam worker and who probably knowing that my father is dead and was even rendered before death long incapable by old age and disease from entering into any public controversy with him sets up a claim to the merit and has the modesty to ask a remuneration from the public for doing in 1811 exactly what I have just now shown your Lordship originated with my father in 1787. . [24]

Harvey and Downs-Rose are however somewhat unsympathetic to Bell and there is in fact no reason why Bell, who had involved himself in the issue of steam navigation and already had possible contacts with Fulton, for example through Rennie, should not have met and discussed these matters of common interest with the American.

Indeed Patrick Miller Jnr cannot be considered the most reliable of witnesses. His object in writing to Melville was to seek to prevent any public reward going to Bell and to state the claims of his deceased father and, himself as heir, to any reward.

Fulton, as we have noted, certainly came to Britain in 1787—which might seem to accord with Patrick Miller's argument, but his initial purpose was not engineering research but rather to study and practice art. Indeed his passage to London was, it is said, sponsored by Philadelphia merchants eager to raise the cultural tone of the Pennsylvanian city. His return to the United States of America, with a Boulton and Watt engine for the steamboat he and Livingston intended to build to exploit the New York monopoly did not take place until late in 1806, which seems rather a long time to have been learning the principles of steam navigation from Patrick Miller Snr.

Fulton's (and America's) first successful steamboat had its trials on the North River in New York State in August 1807 and went into commercial service between New York and the state capital at Albany in September 1807. This was, at 150 miles each way, a considerably more ambitious route than that which the *Comet* was to pioneer in Scotland five years later.

Fulton was to build a number of other steamboats for service on the Ohio and Mississippi rivers and also the first steam powered warship—the *Demologos* or *Fulton*—a heavily gunned, armoured steamship intended for use against the Royal Navy in the war of 1812–14 between Britain and the United States. This is the vessel Bell refers to as

   . . the steam frigate, called *Fulton the First*

in his *Caledonian Mercury* letter in 1816. The huge (at 167 foot long on deck and 2745 tons displacement heavier than, though roughly equal in size to, HMS *Victory*) twin hulled *Demologos* underwent successful sea trials in late 1814. However she was never put to active service and when peace was signed between Britain and the USA in December 1814 *Demologos* was laid up, and was later accidently destroyed, in the Brooklyn Navy Yard. Fulton had earlier tried to interest the US government in his schemes for underwater warfare—with little more success than his experiences with the French and British governments. The twin-hulled construction of *Demologos* may owe something to Fulton's contact with Miller or his writings, it being remembered that Patrick Miller was a great advocate of the multi-hulled ship.

Sadly no correspondence between Bell and Fulton can be traced in any of the American archives which hold Fulton's papers and Fulton does not appear to have made any public acknowledgement of any part that Bell may have played in the development of his plans. However, as Fulton spent much time, effort and cash in the last years of his short life conducting legal defences of his steamboat patents and monopoly rights it is perhaps not to be expected that he would have admitted to having used any of Bell's ideas or worked along lines suggested by Bell or acknowledged the lessons he undoubtedly must have learned from his inspection of the *Charlotte Dundas* and his meeting with Symington.

So anxious was Fulton to assert the autonomy of his inventive genius that he wrote in 1811 to Lord Stanhope asking him to authenticate a letter which Fulton claimed to have written to Stanhope in 1793 outlining his design for a steamboat. Stanhope is not recorded as falling in with Fulton's wishes as regards this letter, which Fulton wished to use in his litigation. Cynthia Owen Philip in *Robert Fulton: a biography* argues that the document which Fulton sent to England for Lord Stanhope to authenticate was a fabrication of 1811 rather than a genuine product of the young Fulton. At the same time as he wrote to Stanhope asking for his letter to be authenticated he also wrote privately to ask him if any steamboats were operating in Britain or Ireland. It was an important part of Fulton's case against the infringers of the Livingston/Fulton monopoly that he alone was the only person anywhere in the world presently capable of designing and bringing into practical service a steamboat.[25]

It will be apparent that Fulton and Bell had much in common, not least a certain propensity to overstate their claims and, it would seem, a tendency to a creative reinterpretation of historical evidence. They also claimed to share a belief in the vital importance of free-trade—Bell was to speak of the function of the steamboat in breaking up monopolies and establishing free trade between all the nations of the earth.

> Wherever there is a river of four feet in depth of water throughout the world, there will speedily be a steamboat. They will go over the seas to Egypt, to India, to China, to

America, Canada, Australia, everywhere, and they will never
be forgotten among the nations.[26]

Fulton had argued, somewhat less convincingly, to the French
government that his submarine *Nautilus* would ensure the
freedom of the seas and thus promote universal free trade. Free
trade was indeed a favourite theme of Fulton's. His belief in this
great principle did not however seem to stop him fighting with
considerable enthusiasm to protect his own monopoly rights.

CHAPTER 7
Steam Boat Proprietor

It is difficult to say with any great assurance exactly when Henry Bell first turned his mind to the possibilities of steam navigation. It may have been while serving his apprenticeship as a millwright that he decided to leave the family business and venture into this field or the idea may have taken more precise shape during the period of his employment, at the age of nineteen, in the Bo'ness shipyard of Shaw and Hart.

In Bo'ness he would certainly have had excellent opportunities to see, in close proximity, the traditional skills of sailing and shipbuilding and the new technology of the steam engine applied to mines. It is quite conceivable that these two technologies may have converged in Bell's imagination. As was discussed in Chapter Four the search for a satisfactory form of mechanical propulsion for ships was a topic which had for long engaged many minds throughout the Western world.

Even if the idea of steam propulsion had not fully formed in Bell's mind during his Bo'ness period then the reports of Patrick Miller's first experiments, carried out in the Forth in the year after he left Shaw and Hart's yard, may well have given him some fruitful ideas. Certainly his experiences in London with Rennie, working for an engineer at the centre of new technological thought, must have been a formative experience. William Bain suggests that Bell had produced a model of an engine for a steamboat as early as 1799.[1]

John Thomson gives one view on the origins of Bell's ideas on steam power. Thomson was a Glasgow engineer who was employed by Bell to carry out ship model trials prior to the commissioning of the *Comet*. It must be said from the outset that Thomson is an undoubtedly hostile witness. He was convinced that Bell had stolen his work and he gave his version of events to the world in 1813 in a pamphlet entitled *Account of a Series of Experiments, made for the Purpose of Ascertaining the Best Mode of Constructing Vessels with Machinery, to be Wrought on Navigable Rivers by the Power of Steam. Including a Review of a Pamphlet, lately Published, intitled, 'Observations on the Utility of Applying Steam Engines to Vessels, etc By Henry Bell'.*

Thomson first met Bell in 1811 and the pamphlet claims that Bell's original ideas, as communicated to him that year, were for a river passage-boat to be worked by hand-cranked machinery.[2] Such a retrograde proposal would, of course, have gone back beyond the *Charlotte Dundas* and the Dalswinton steamboat to the hand-propelled experimental boat which Patrick Miller had sailed in the Firth of Forth in the Spring of 1787.

Bell himself says in his *Observations*. . .

> In the year 1809 I attempted to make a small model, in which I succeeded so far, that I was convinced an engine could be made on such a construction so as to drive a vessel in all weathers. But for further information before I attempted anything on a large scale, in the year 1810 I built a small boat 13' x 5' in which I tried a great many experiments, by erecting a number of different machines; and in the year 1811, I succeeded so far in my views, that I was fully convinced a vessel of any size could be wrought by steam.[3]

However, of these experiments Thomson notes:

> I have made enquiry, since the publication of the Obser-vations, concerning the nature of these experiments, as I never had heard any thing of such being made, either from

Mr Bell himself, during our acquaintance, or from any other person; but my enquiry has been fruitless.[4]

Of course Bell was later to go on record with a claim that he had been applying his mind to steam navigation since at least the turn of the century and would repeatedly cite in support of this contention his supposed Admiralty applications of 1800 and 1803. However, as has been demonstrated in the last chapter, the weight of the evidence suggests that these applications must be considered as solely the product of Bell's imagination. On the other hand Edward Morris states that it was indeed while Bell was working with Shaw and Hart at Bo'ness in 1786 that:

> ... he formed a strong—a lasting, and as it turned out, a triumphant impression of the power and applicability of steam to transmarine purposes; . . .[5]

and Morris later on cites Bell's own lost memoirs in support of this view:

> It appears from Mr Bell's own writings that it was at Borrowstowness, in 1786, when with the Messrs. Shaw, that his mind was strongly impressed with the steam-boat system. He had a present made him, when a boy, of a small rigged sloop. To this sloop he paid much attention, to the exclusion often of his school education. When he had served out his apprenticeship, in the mill-wright line, he was often ruminating on the little sloop, which had charmed him in his boyish days, in sailing it on the streams, and catching the gentle breeze to aid its onward course. But the stream would often drive it in a direction contrary to his wish; and the winds, either still or adverse, would often cause him to reflect on the possibility of steam to counter these opposing forces:— here was the *Comet* in embryo.[6]

As has been suggested before it is always difficult with Bell to distinguish the later inventions and *post-hoc* rationalisations from

136

the unvarnished truth. In fairness it must be said that his 1813 pamphlet is, as was demonstrated in the last chapter, a more modest and credible statement of his achievement than some of the later writings. Indeed as it was written so close to the events it narrates there would have been many readers of his pamphlet with first hand knowledge and a clear recollection of the events and correspondingly little opportunity to embroider the truth.

When he gave his more colourful version towards the end of his life he was writing fourteen years after the *Comet*'s maiden voyage and could perhaps rely on the passage of time to obscure the truth or falsehood of some of his claims. However the 1813 pamphlet may be viewed with more confidence and there may be considered to be no real problem in accepting his account of experiments in 1810 and 1811. One may also cite the correspondence with James Watt on Watt's views on steam navigation as evidence in support of the view that his mind had long been exercised by this idea. This exchange of letters is supposed to be dated around 1801. However this correspondence seems only to exist in reported form and in the absence of better evidence this may, perhaps, be yet another piece of Bell fiction.

Despite Thomson's claim that it was he who had awakened Bell to the possibilities of steam power, there seems little doubt that Bell had genuinely been giving thought, in however unfocused a way, to mechanical propulsion for many years prior to the building of the *Comet*. It is quite possible that he did not take John Thomson fully into his confidence and it must be equally possible that the man-powered experiments Thomson was engaged to conduct were intended by Bell as no more than economical ways of testing drive mechanisms, paddle arrangements and paddle designs. It seems most improbable that anyone, whatever his deficiencies in technical knowledge might have been, who had lived through the period of Patrick Miller's capstan driven experiments and Miller's rejection of man-power in favour of steam-power and who had, on the evidence of James Taylor, seen and had demonstrated to him the steam powered *Charlotte Dundas* would have invested time and money in such a limited and backward-looking scheme. We also have the suggestion given by Taylor in his letter to William Huskisson, President of the Board of Trade, in 1825 that:

> ... when Mr Bell was at any loss, he paid visit after visit to lock 16 to get round his difficulty.[7]

Lock 16 at Camelon, on the outskirts of Falkirk, on the Forth and Clyde Canal was, of course, where the *Charlotte Dundas* was laid up following the Canal Company's withdrawal of support and ban on steam propulsion. It is more than a little difficult to square this story of Bell paying repeated visits to the *Charlotte Dundas* with Thomson's account of Bell's first thoughts being only of man-powered propulsion—the most extreme version of which he gives in the following colourful description:

> ... he [Bell], in the most positive manner, declined being at any farther expence; adding, that it was of no use to him, if a stout boy, by the application of mechanical powers, could not row a boat from Glasgow to Helensburgh with ten or twelve people on board. This I knew, and stated to be impossible, as a man could apply his whole strength to an oar with equal propriety as to any other machine whatever.[8]

There is however another yet account of the development of Bell's thinking on this topic. The Rev William C Maugham in his *Annals of Garelochside* . . . (1897) suggests that Bell had seen an earlier hand-powered paddle-wheel driven wherry, owned by Andrew Rennie of Greenock, and that his knowledge of this vessel, known as "Rennie's wheel boat", had resulted in him ordering his model boat from Nicol's shipyard in Greenock[9], a noted builder of the Clyde fly-boats and other small vessels. Maugham suggests, and to this extent his story accords with Thomson's version, that at first this experimental model operated as a stern-wheeler but that side-mounted paddles were later fitted. However no great speed was attained by this model and Henry Bell was forced to conclude that only steam power would suffice. This version, whose source is not identified, has a certain attractiveness with its specific reference to Andrew Rennie's craft, but does not satisfactorily answer the point that Bell, prior to the year 1810, knew all about the steam powered *Charlotte Dundas* and had met and discussed her and the whole subject of steam navigation with Robert Fulton.

We are perhaps forced to conclude that Bell was not being totally frank with Thomson. If so, then he would not be, by a long way, the first inventor to keep his own counsel and to hold matters back even from his most trusted collaborators. Thomson had after all only come into Bell's life in 1811, when he came for a short visit to the Baths Inn to take the health-giving sea air of Helensburgh and, while he was in residence there, struck up an acquaintance with Bell. During Thomson's stay in Helensburgh he had many conversations on technical matters with his landlord. He then returned to Glasgow where Bell called on him and contracted with him to carry out work on his model boat which, by Bell's account, had already been built or, if one follows Thomson's version, was even then building.

Thomson states that, after the successful trials he conducted on Bell's behalf, Bell proposed that they enter into partnership to develop the steamship idea. He tells how the two men divided the Clyde between them in the search for financial backers for this project. However Thomson, despite considerable circumstantial detail elsewhere in his pamphlet, somewhat obscurely accounts for the failure of this prospective partnership to reach fruition in the rather anodyne words:

> Various circumstances occurred to prevent this plan from being effected. . .[10]

It is certainly not out of keeping with what we know of Bell's character to believe that he was more than capable of deliberately misleading Thomson and hiding his true intentions from him. The tale of his elaborate survey for the Glen Fruin reservoir and his teasing of the farmer (*vide supra* Chapter Five) indicates a lively and unorthodox nature and taken together with the obituary comment ". . . his ruling passion, scheming . . . " suggests a man who could not necessarily be counted on to take the straightforward course in any matter.

We have in any case the notes of John Buchanan's interview with John Robertson, the builder of the *Comet*'s engine, which state that:

> As far back as 1807, Robertson had frequent conversations with Bell, about the practicability of propelling a boat by machinery.

Admittedly this statement does not positively identify the proposed machinery as being steam-powered, rather than man-powered. However, as man-powered propulsion would, from the perspective of Buchanan's conversation with Robertson in 1853, have so obviously been a blind alley on the way to the achievement of the *Comet* that it would be reasonable, if such a notion had been a serious part of Bell's deliberations, to have expected some mention to be made of it by John Robertson. In addition, as Robertson went on to say that they:

> Both had seen Symington's vessel lying in one of the reaches of the great canal, and the idea was taken from her.[11]

one may be reasonably confident that it was steam-power which formed the subject of the two men's conversations about mechanical propulsion.

The rights and wrongs of Bell and Thomson's working relationship are somewhat obscure. It is difficult to decide how much credit is due to Thomson and how far his ideas were unfairly appropriated by Bell. Not all Thomson's assertions can be accepted at face value and his *Account* has a number of internal inconsistencies. For example he says that he and Bell in their discussions touched on:

... the working of vessels by the power of steam, knowing, at the same time, that they were used to a great extent in the rivers of the United States of America.[12]

Bell must have been well aware of the use of steam on American rivers from his meetings with, and probable correspondence with, Robert Fulton. Even if Bell had not maintained a correspondence with Fulton there seems no reason to doubt that he would, if only from the fact of having met him, have had an interest in and knowledge of his well-recorded achivements. It thus seems most unlikely that Thomson, though by his own account an experienced erector of steam engines, would be able to tell Bell anything he did not already know about the American situation. If, as Thomson states, he and Bell discussed American steamboats, and even more if Bell knew all about Fulton's North River steamboat then it is surely even more unlikely that he was genuinely planning a hand-cranked passenger boat for the Clyde. Thomson nonetheless later suggests that it was only his advocacy of steam power that persuaded Bell to abandon his man-powered ideas.

Thomson's opinion was that his experimental work on the model boat, which he claimed had culminated in the offer of a partnership and in Thomson preparing the technical drawings of the ship and machinery, had, despite the abandonment of the partnership, entitled him to win the contract for supplying the *Comet*'s machinery. In Bell's defence it could be argued that Thomson was simply a technical expert hired to do a job and that his rights and interests in the matter did not necessarily extend beyond that. This case would argue that Bell had no obligations to him other than to pay him for the work done and for services rendered. In the light of Bell's failure to pay Robertson and Napier and Wood it should be observed that Thomson was paid for the work he had done on the model boat:

... Mr Bell called to settle the expence incurred in the experiments done by his orders, which amounted to something more than £23.[13]

141

There does however, it must be admitted, remain something of an air of unpleasantly sharp practice about the idea, if true, of Bell taking Thomson's drawings and using them to commission his chosen contractors in Glasgow. One must however treat Thomson's comments with some reserve. There is throughout Thomson's *Account* an unmistakeable tone of bitterness and he consistently presents all Bell's actions in the worst possible light. Even when Thomson feels constrained to praise Bell he manages to do so with more than a faint damn.

> I certainly do not say but that the After-cabin of this vessel [the *Comet*] is fitted out in a very tasteful manner; and Mr Bell, or whoever directed it, is justly entitled to every commendation; at the same time I would certainly have preferred making the space for the passengers larger. . .[14]

John Thomson's qualifications to comment on steamboat design are not in doubt. He was himself to own and to build the engine for the Clyde's second steamboat, the *Elizabeth*. This vessel at 58' overall, 51' on the keel, was considerably larger than the *Comet* and was placed in opposition to her on the Glasgow to Greenock route with some success. It was in fact the opposition provided by larger and faster vessels such as the *Elizabeth* that was to lead Bell to take the *Comet* off the Clyde in search of more profitable trade elsewhere. As Thomson reported in his *Account* the *Elizabeth* had:

> . . . sailed 81 miles, in one day, making an average of actual sailing not short of nine miles per hour.

Thomson, it is interesting to note, despite his differences with Bell, had somewhat similar ideas on the rather domestic style in which steamboats should be fitted up, and his *Account* goes on to describe the appointments of the *Elizabeth*'s best cabin:

> A sofa, clothed with marone, is placed at one end of the cabin, and gives the whole a warm and cheerful appearance. There

are twelve small windows, each furnished with marone curtains, with tassels, fringes and velvet cornices, ornamented
with gilt ornaments . . . at each side book-shelves are placed,
containing a collection of the best authors, for the amusement
and edification of those who may avail themselves during
the passage.[15]

Marone is simply a eighteenth and nineteenth century spelling
of maroon—*Elizabeth*'s upholstery and curtains were thus of a
deep brownish-red colour.

Whatever the preliminaries and whatever stages Bell had
to come through to arrive at his grand design it was towards the
end of 1811 that he arranged for John Wood to build a steamboat
for him. The design of the *Comet* and her machinery probably
owed as much to the example of the *Charlotte Dundas* (although
that vessel's stern-wheel arrangement was abandoned) and to
the benefits of Bell's discussions over the years with Fulton and
later with John Robertson as it did to the experimental work
carried out by Thomson.

The size and design of the *Comet* has already been discussed
in the opening chapter. There is however at this stage one point
of real significance which should be made and which is of critical
importance in any consideration of Bell's career as a ship-owner.
So far as one can judge, Bell's ambitions for steam navigation in
general, and for the *Comet* in particular, were extremely wide-
ranging and ambitious. In this respect he differed from such
earlier pioneers as Patrick Miller. Miller had become involved in
the question of the mechanical propulsion of vessels through
the public spirited intention of providing an auxiliary means of
propulsion which could be employed to save life at sea. James
Taylor records his employer's laudable, if limited, ambitions thus:

> I will explain my views—my object is to add mechanical aid,
> to the natural force of the wind; to enable vessels to avoid,
> or extricate themselves from dangerous situations, when they
> cannot do it on their present construction; and I wish also to
> give them powers of motion in times of calm.[16]

Taylor goes on to say that every report of shipwreck or loss provoked in Mr Miller an outpouring of philanthropic feeling and a renewed determination to push forward with his experiments. The successful culmination of Miller's experiments came with the trials on the Forth and Clyde Canal.

John Thomson's views on the uses of steam power were, if we may fairly judge on the evidence he gives us in his *Account*, equally, if differently, limited. The middle portion of the extremely lengthy title of his pamphlet refers to "...*Constructing Vessels with Machinery, to be Wrought on Navigable Rivers*...". Thomson with the *Elizabeth* had produced an effective steamboat but his ambition for it was limited to service on "Navigable Rivers" such as the Clyde.

Bell, on the evidence of his *Observations* and judging by the standard of equipment of the *Comet*, beds in the cabin and so forth, always seemed to have had wider aims and greater vision. In his *Observations* he points out the applicability of steam engines to ferries plying on open-sea routes such as that from Scotland to Portpatrick in Northern Ireland or the Queensferry passage across the Firth of Forth.

At the time of publishing his *Observations* even the fervent Bell did not publicly advocate the exclusive use of engines on ocean going ships (which was not indeed to come into practice for many decades) but he did enthusiastically promote their use in an auxiliary role in such circumstances. That said, it may well be that Bell did see much further and more imaginatively than he thought politic to admit to publicly in February 1813.

Edward Morris has a number of accounts of Bell's vision of a world linked by the steamship. On one occasion he reports Bell as arguing in favour of free trade against an advocate of protectionism:

> Sir, my steamers will break down all your schemes—they will fight their way against all monopolisers, and you *cannot arrest their course!*[17]

And Morris also relates how:

At a public dinner given to a number of gentlemen in the Baths Hotel, Helensburgh, some years ago, Mr Bell was speculating in an animated tone on the probability of steam ships going to every land, and carrying our merchandise over the wide Atlantic, the Indian and China seas—to *all* nations. Several gentlemen laughed at his *"wild ideas,"* as they termed them; but Bell spoke rationally; he saw deeper into things than the *laughers* did, and his opinion has been gloriously realised.[18]

Henry Bell's breadth of vision and imaginative grasp of the great potential of steam power did not just extend to the steamship. He was evidently keenly interested in all aspects of steam powered transport. This again gives a persuasive indication that John Thomson's suggestion, that it was he alone who had persuaded Bell to adopt steam propulsion, can have little validity. Bell's vision of the new age of steam power is given in a lively narrative by Gabriel MacLeod as told to his nephew Donald MacLeod, the author of *A Nonogenarian's Reminiscences of Garelochside and Helensburgh.*

Henry Bell used betimes to go into my brother Donald's joint tailoring and dram shop to have a crack with him. The performances of the *Comet* steamer being with them one day the theme of conversation, my brother said as a fitting wind up thereto, 'Man, Mr Bell ye're a desperate clever chiel, that boat o' yours is just a perfect world's wonder.' To which the ardent-minded proprietor replied, 'Danney, tak' my word for it this is only the beginning of the uses that steam engines will be put to in the way o' conveyin' passengers; if ye leeve lang ye'll see them fleein' and bizzin' about on land, wi' croods o' passengers at their tail, lively as a spittle loupin' alang a tailor's het "goose".[19]

Quite apart from its value as an indication of Bell's thinking this account in a lively Scots dialect is of interest as being likely to be a rather more authentic example of Bell's conversational style and vocabulary, than Edward Morris's genteel and sanitised version of his words rendered into a polite English. To be fair to Morris he indeed writes that there was:

> ... a richness and a genius in his thoughts and communications, but the rough garb in which these were arrayed was to his disadvantage, especially with those persons (and there are too many of them in the world) who look more on the surface than into the depths of things.[20]

But throughout his biography he edits Bell's writing and speech into an unpersuasive conformity with standard English.

However expressed, Bell's visions of the new age of steam were, in the eventful year of 1812, obliged to take second place to the hard practicalities of making the *Comet* a success. Indeed even getting *Comet* launched proved not to be without its share of problems. One reason for the long delay from a contract date in late 1811 to a launch date in the month of July or August 1812 would appear to have been Bell's shortage of funds. John Robertson reported that the *Comet* lay, completed, in John Wood's yard but that the builder refused to let her be launched until he was paid. Robertson's recollection of the events was that the situation was only solved by Bell finding a guarantor, whom he believed to be a Mr Kibble of Dalmonach. This was most probably James Kibble, a partner in the Dalmonach Print works in the Vale of Leven, Dumbartonshire. The Dalmonach Works had been destroyed by a fire in 1812 and were being rebuilt under Bell's superintendence even as the *Comet* was taking shape at Port Glasgow.

Such far from unusual awkwardnesses over Bell's liquidity being satisfactorily resolved, *Comet* duly took to the river and plied her intended route from Glasgow to Greenock on Tuesdays, Thursdays and Saturdays. Sunday, of course, being then strictly observed as a day of rest from all but essential work, the return trips up-river were accomplished on Mondays, Wednesdays and Fridays. One of the motives behind Bell's scheme was his desire to find a comfortable and convenient way to bring his hotel clients down to the Baths Inn. The "Lady Charlotte" sociable and the "Mermaid" light coach which had served the town and had brought Bell's early clients to the Baths Inn had been neither swift nor comfortable. However in the early months of the *Comet's* operation, and despite the obvious wish to bring visitors down in comfort, the connection to Helensburgh was at first normally accomplished by transferring passengers to a wherry, a small shallow draft vessel, at Greenock. The wherry then made the three mile crossing to Helensburgh by the old and tried power of wind and oars. Such ferries might have been tried and tested but were by no means always a safe means of transport, as was to be demonstrated in February 1814 when a boat on an adjacent cross river service, the Port Glasgow to Cardross ferry, sank with the loss of eight lives. The use of the ferry as the last link in the route does tend to suggest that any explanation of Bell's motives for investing in steam navigation which centres exclusively on the *Comet* as an adjunct to his hotel business is an over-simplification. Again one is forced to conclude that Bell always had wider aims and a more imaginative vision.

An exception to the *Comet's* normal schedule came on her Saturday sailing when, according to John Robertson, after calling at Greenock, she sailed to her home port of Helensburgh and remained there over the weekend. Bell's motives for this routing would appear to have been a mixture of a wish to avoid the harbour dues he would have had to pay had *Comet* lain at Greenock and the desire to have the ship to hand so that he could supervise its maintenance and any necessary repairs.

One slight difficulty for Bell and the *Comet* was the lack of any satisfactory pier at Helensburgh. The lack of satisfactory harbour facilities was to be a long-standing grievance in the town and some twenty-five years after the *Comet* started sailing the well known artist Joseph Swan in an engraving entitled "Helensburgh, from beyond the Baths" depicted a sailing vessel somewhat precariously beached at a rocky spit close to the Baths Inn. It was presumably at this same spot that *Comet* or the

147

wherries from Greenock landed their Helensburgh-bound passengers.

There was, at first, some fairly natural resistance to the new means of transport. Indeed a degree of hostility seems to have existed in some quarters. Even seven years later, when Bell was extending the *Comet*'s sailings to the West Highlands, an Argyllshire clergyman was to write:

> . . . In Helensburgh . . . there dwelt a man of skill in the art of the Engineer. He being puffed up in pride of his Abilities therein did lately Conceive a Ship to go on the waters, not by dint of the Clean winds of the Air as ordained by God, but by means of fire which burned and a great Smoke which issued from the bowels thereof. In this Device the Man did sail upon the waters of the River Clyde in Pride and highness of Heart, to the great scandal of pious minds; but also to the Huge Gratification of the Man and those that were with him. But, not Content to keep his Unholy devices in the scene of their Conception, this Man must needs sail or propel his Creature even unto Crinan itself, near to which place your Servant labours in the Gospel both in The Gaelic and English. And so to Fort-William, with much effusion of stinking vapour and great hurt to the light-minded of this Congregation and of the Christian Consciences of the People of Argyll, The Isles, and even Inverness.[21]

However before long it became apparent that the steamship was, despite its trail of smoke and sparks from the tall funnel, a safe and reasonably reliable means of transport. In a remarkably short time most people with occasion to travel between Glasgow and the coastal towns, with the exception of such enlightened thinkers as the clergyman quoted above, came to make use of the steamer and the fly-boats and stage coaches fell into some decline. A technology having been established and a market having been created the *Comet* was quickly joined on the river by rivals. In February 1813 John Thomson's *Elizabeth* commenced a daily service—an obvious improvement on the *Comet*'s thrice weekly schedule. The new steamer charged five shillings for the best cabin—against Bell's fare of four shillings—but otherwise undercut *Comet* by one shilling with a second cabin fare of only two shillings. The daily service, departing from the Broomielaw at 8.00am was advertised as:

> . . . safe, pleasant, expeditious and cheap.[22]

By June the *Elizabeth* had extended her route to Gourock—a popular watering place and port a couple of miles beyond Greenock and had adjusted her fares, the Greenock fares now being four shillings and two shillings and sixpence. These changes to *Elizabeth*'s schedule and fares were probably largely influenced by the prospect of even more competition when the third steamer on the river, the *Clyde*, came into service. This event took place in July 1813 and she advertised daily sailings to Greenock and a thrice weekly continuation to Gourock. A stop at Port Glasgow was also offered. The *Clyde*'s owners were able to boast that:

> The *Clyde* being by far the largest boat of the kind in the river, has excellent accommodation for Passengers. The fire and machinery being completely enclosed, the Passengers are neither incommoded with the heat of the Fire, or the noise of the Engine; and, in this respect the *Clyde* is superior to any other Boat of the kind in Scotland.[23]

By April of the next year, 1814, the *Glasgow* steamer was in service and the *Trusty* steamer was announcing the commencement of her thrice weekly sailings between Glasgow and Greenock. The proprietors of the *Trusty* advised readers that their ship had made the passage in 2 hours 48 minutes and, as importantly, they undercut the competition by offering cabin fares of only three shillings.[24] However *Trusty* was to become within a few years principally a cargo, rather than a passenger steamer, presumably through being forced out of the passenger trade by the advent of

THE INGENIOUS MR BELL

ever larger, faster and better equipped ships coming into service.

Perhaps concerned by this increasing competition the proprietors of the *Clyde* took an advertisement in May 1814 to remind readers of the *Glasgow Courier* that their ship was the largest on the river and that no expense would be spared by them to ensure complete public satisfaction.[25] Not all went smoothly even for the excellent *Clyde* and the hazards of the sea were experienced by her and her passengers in full measure. A story in the *Glasgow Courier* in November 1814 which from its sympathetic account of the accident, its emphasis on the passengers loyalty and the owners' exertions on their behalf, seems to have some of the hallmarks of a planted story, or what we would now think of as a company press release. It told how:

> . . . the *Clyde* Steam Boat, owing to a great fog, struck on the long-dyke a little below Dunglass. The majority of the passengers wished to remain in the steam boat rather than go in the two boats which the owners of the *Clyde* had provided for them; some time after, she began to sink by the stern,owing to the water falling away, when they all got safely on shore in two boats belonging to a brig lying in the river.[26]

The long-dyke the *Clyde* struck was part of Thomas Telford's scheme to deepen the Clyde by building parallel banks to increase the scouring effect of the river's flow. In the following year the dangers of working in the engine room of the steamers was brought home when the *Clyde's* engine man had to have his arm amputated after it had been mangled by the machinery.

By 1815 the River Clyde was becoming well-served by steamboats. In addition to those already mentioned the river towns now received regular calls from the *Prince of Orange, Princess Charlotte, Caledonia, Dumbarton Castle, Britannia* and *Greenock* steamboats. In addition to the original services to Greenock and Gourock the other towns on the Clyde or its tributaries soon had their own services. In July 1815 the *Prince of Orange* was the first steamer to make the difficult three mile journey up the shallow River Cart to call at the important textile town of Paisley. This auspicious event was witnessed by large crowds who turned out on the quay to see Paisley's integration into the network of Clyde steamer services.

As important for the success of the steamboat as the ever-widening range of service was the growing acceptance of the safety and convenience of the steamer. Press reports, such as this from 1815, must have been very helpful to all the steamer owners in developing public confidence in the new mode of travel.

> The *Greenock* Steam Packet, which sailed from Glasgow on Friday last for Inveraray did not, owing to the late boisterous weather, return to the Broomielaw until yesterday afternoon. Off Otter Ferry she encountered a dreadful storm, and at Tarbet she was drove from her anchors, which were, next day, recovered; but by her powerful machinery she was soon carried into Tarbet Harbour. Two naval officers, who were passengers, are of opinion that scarcely a cutter in his Majesty's service, could have rode out such a gale as the Greenock encountered.[27]

With the increasing degree of competition on the river Bell had early on been placed in the unenviable, but by no means unusual position, of being the pioneer who failed to profit by his pioneering. In 1813 his financial position was such that several of the prominent citizens of Glasgow and district, including Kirkman Finlay, the city's Member of Parliament, and the Members of Parliament for Lanarkshire and Dumbartonshire, aware that Bell had spent more on his experiments than his limited means could well afford prevailed on him to accept assistance from them towards defraying part of his outlay.[28]

Bell himself described, in a letter published in 1825, the early history of his career as the owner of Europe's first steamship company:

> In 1811, I built the *Comet*, and plied her between Glasgow and Greenock; but the prejudice was so great against steam

boat navigation, by the hue and cry of the fly-boat and coach proprietors, that for the first six months very few would venture their precious lives in her; but in the course of the winter of 1812, as she had plied all the year, she began to gain credit, as passengers were carried 24 miles as quick as by the coaches, and at two-thirds less expense, besides being warm and comfortable. But even after all, I was a great loser that year. In the second year, I made her a jaunting-boat all over the coasts of England, Ireland and Scotland, to show the public the advantages of steam boat navigation, over the other mode of sailing.

One should note that Bell has yet again managed to get the wrong date for the *Comet's* entry into service by saying that she was built in 1811. As a result he suggests that she traded all during the year of 1812 rather than only from August of that year. He goes on to say, and here he is looking back over his whole ship-owning career, but the comment has equal validity for the early period on the Clyde:

> ... I experienced a great loss in opening up all the different grounds single handed, which, as soon as they were seen to pay, induced large and powerful companies to embark in steam boat speculation with larger boats and greater power.[29]

To be fair, Thomson's *Elizabeth* was hardly the product of a "large and powerful" company. John Thomson was basically a blacksmith from the Gorbals, despite his self-description as an engineer—much as Bell was a wright from the Gorbals who at times described himself as as an architect. The essential point is that with the increasingly frequent arrival on the river of a number of larger and faster rivals Bell was probably wise to look for other areas of trade.

It was while the *Comet* was, in Bell's phrase, a "jaunting-boat" that he took her to the Forth. This journey must have involved a delicate negotiation with the Forth and Clyde Canal Company, to allow the passage of a steamboat through the canal.

The Company had, it will be remembered, banned steamboats from their waters after the *Charlotte Dundas* experiment for fear of damage to the banks of the canal from the wash of the paddle-wheels. Bell had two objects in mind for his trip to the Forth, the first as he says to "show the public the advantages of steamboat navigation" and perhaps to open up a new route on the Forth; the second, to have a refit carried out under George Hart's superintendence at his old employer's Shaw & Hart's yard at Bo'ness.

While *Comet* was at Bo'ness a steamboat excursion was run—apparently the first of its kind on the Forth—from Bo'ness to Leith at a single fare of seven shillings and sixpence. On 24th May 1813 the *Edinburgh Evening Courant* carried a one sentence report:

> The *Comet* of Helensburgh, a vessel worked by steam, and the first of the kind ever seen in this quarter, is at present lying in Leith harbour.[30]

On this, or another of his pioneering visits to Leith, Bell ran trips in *Comet* and took the opportunity to show her off to the Edinburgh population. We get, even through the excesses of bitterness and sarcasm, a good description of Bell's promotional activities for the *Comet* from the sharp pen of John Thomson:

> It is not merely in his Observations that Mr Bell has arrogated to himself the merit of this invention. In a late trip which he made to Leith with the *Comet* Steam Boat, he lost no opportunity of gratifying his personal vanity in this way. The public Newspapers also announced it in various shapes. This wonderful vessel was advertised to be exhibited in Leith Harbour, (something in the manner of an itinerant's show-box), for the inspection of the citizens of Edinburgh—and the accommodating, public-spirited, disinterested would-be inventor, was even willing to give the good folks of Edinburgh an opportunity of being carried out to the Roads, *by the power of Steam*, for a small pecuniary equivalent.

> Whether such a method was likely to impress the public with
> a high opinion of Mr Bell's talents, or the importance of his
> assumed invention, I leave to them, as well as Mr Bell himself,
> on mature consideration, to decide.[31]

It is hard not to feel that Bell was displaying a fairly high standard
of marketing and entrepreunerial skills in such activities. It is
difficult to sympathise with Thomson's apparent view that for
some reason or another the steamship was of such importance
that it should not be exposed to any such commercial exploitation. Quite why Thomson, a steamship owner presumably
concerned to make a financial success of his *Elizabeth*, should
have taken such a oddly perverse view is unclear. One feels that
the most likely explanation for his view is that the extremity of
his bitterness against Bell warped his judgement.

When *Comet* had first entered service she had been
commanded by Captain William MacKenzie, who, due to ill
health, resigned his position at the end of 1812. The name of
MacKenzie's immediate successor does not seem to have come
down to us but the skipper most famously associated with the
Comet, Captain Robert Bain, took over command in 1814, and
was to serve throughout the ship's life. In the light of the various
reflections on Bell's character it is just to note that Bain and Bell
would seem to have worked well together, Bain being later on a
shareholder in the *Comet*. When Captain Bain died at the age of
39, having spent most of his working life in Bell's employment,
Bell was to erect to his memory, in Rhu churchyard, a large iron
tombstone, the product of the Shotts Iron Works, which outlines
Bain's career and was as the inscription states:

> . . . erected as a tribute of honour for 16 years faithful service
> by Henry Bell, Engineer, Helensburgh.

It was not long before the steamboats were doing more than
simply providing a ferry service between Glasgow and the
commercial ports of the Firth of Clyde like Greenock and Port

Alexander Nasmyth's painting of the Comet *on the River Forth records Bell's attempts to open up other areas to steamer traffic. (Science Museum)*

Glasgow. In September 1814, for example, the *Glasgow* steamboat was advertised to carry out what looks very much like a weekend pleasure excursion—departing from the city on Saturday at 9.30am for Greenock, Gourock and the Ayrshire resort of Largs; returning from there, after the obligatory Sabbath day of rest, on Monday morning.[32]

During the summer of 1815 on some days the Glasgow press carried advertisements for three or four different steamers offering new routes and special excursions to Rothesay and various locations in Argyllshire as well as day or weekend trips run to connect with special events such as the Largs Fair or the Marymass Races at Irvine.

The early steamers were based in Glasgow and sailed downriver from there each day, but now new steamers were being brought into service offering the complementary routing, from the country towns into Glasgow. One example of this was the Dumbarton based *Duke of Wellington* steamboat which commenced a daily service to Glasgow from Dumbarton quay on the River Leven in September 1815.[33] Her locally based management made a point of advertising that post chaises would be in attendance at Dumbarton to take passengers on to Helensburgh and to Loch Lomondside and other local beauty spots.

By July 1816 the steamboat traffic on the Clyde had developed to such an extent that the various owners found it worthwhile to band together in a cartel to establish uniform price levels.

Ironically enough, Helensburgh, the home of Henry Bell and the target for the first steamer service, was now becoming an area of keen competition with services offered by his rivals, both through connecting road services, as in the case of the *Duke of Wellington,* and directly with steamers such as the *Britannia* calling there.

Bell, just like the owners of the steamboats such as the *Greenock* and the *Glasgow* who promoted sailings to the beauty spots of Loch Fyne and the Kyles of Bute, was always on the look out for new markets for his ship. His comment that he had opened up the various areas to steam navigation is certainly justified and the story of his attempts to exploit the Hebridean and West Highland market will provide an excellent example of this aspect of his career. However even before his involvement in the West Highland adventure he was looking around, closer to home, for likely opportunities for the profitable employment of steamboats and in February 1817 he had a long and detailed letter published in the *Scots Magazine* on a "Plan of Communication by Steam Boats between Leith and Greenock".

This letter is of interest on several counts, not least for its visionary quality and characteristic self-assertiveness which clearly foresees the triumph of steam-power in all coasal shipping:

> Some proof may naturally be expected of my ability to perform what I now propose; and at present, I would merely refer to what I have already effected for the benefit of my country, *viz*, the steam-boats on the Clyde and on different rivers in Europe, which answer the purpose of conveying passengers to any given distance, quicker than mail-coaches; and as I have got the machinery much simplified, I expect in a short time to see all our ferries, and our coasting trade carried on by the aid of steam-vessels.

Such a plan to link Leith and Greenock would of course involve the use of the Forth and Clyde Canal and he gives detailed timings showing that goods could be carried between the two towns in twenty to twenty-four hours:

> The merchant, knowing the time of the tide, can count to an hour, in ordinary weather, when his goods will arrive; and will not be disappointed in one case out of thirty, for six months of the year; and for the other six months in not one case out of twenty, while the Canal and rivers are kept open.

The remark about the Canal being "kept open" is a reference to one of the major problems of such waterways—closure due to ice.

He contrasts this to the existing situation where trading vessels between Greenock and Leith could only make ten round trips a year via the Pentland Firth and merchants' goods were unavoidably tied up for lengthy periods.

His letter recommends the building of four lighters to ply between Leith and Port Dundas—but mindful of the Canal Company's fears for their canal banks he plans to have the canal section of the passage horse towed because only one steamboat is included. His fleet would typically be disposed of thus:

> . . . one loading at Leith, and one at Port-Dundas, one tracking down the Canal, and one towing up from Leith.

The cost of this flotilla was quoted as being £1200 for the lighters and £2000 for the steamboat to take the lighters up and down the Forth from Leith to the Canal sea-lock at Grangemouth. A similar provision of lighters and tug was presumably intended for the Port Dundas—Bowling—Greenock stage.

Bell explains how this proposal would be to everyone's advantage—with additional revenue for the Canal Company, the Greenock and Leith Harbour Trustees, and not least of all benefits for the general public who:

> . . . will have their goods carried from Greenock and Port Dundas to Leith, or from Glasgow to Greenock, at half the expense of land carriage and in fully as short a time.

Obviously such a scheme had particular advantages for bulky and heavy loads such as grain, timber, building materials etc which taxed the ill-made roads and limited land transport facilties of the period.

Bell pointed out how the advent of steamboats on the Clyde had been of great public benefit. In an interesting echo of his remarks quoted above on the opposition he had at first experienced from the coach proprietors he wrote:

> When I first made the steam-boats ply on the river, there
> were seven daily coaches on the road betwixt Glasgow and
> Greenock, of which the boats shortly laid aside five.

He noted that the average fare on the stage-coach was seven
shillings as against a steamboat fare which averaged only three
shillings—and he calculated that there was thus a saving, even
without allowing for the increase in traffic, to the public of no
less than £3120. Bell even goes on to account for the savings in
oats, beans and hay for the horses formerly employed in the
stage-coach trade. In his minutely detailed calculation of the
benefits of the coming of the steamship Bell of course has ignored
the redundant coachmen, stable-lads and horse-breeders to say
nothing of the lost markets for forage for the farmer!

Having set out these plans in some detail—plans which it
seems fairly clear from all we know of his usually precarious
finances that he would not be in a position to implement from
his own resources—he ends by pointing out that such a scheme
would prove advantageous to the shipping companies trading
between Leith and London and says:

> Should any of these Companies wish to embark in this
> speculation, I am fully of opinion that it will answer the
> intended purpose, and not only prove of general utility to
> the public at large, but also profitable to the adventurers.[34]

The implication of the letter is that if some enlightened and
adventurous investor would put up the required capital of £3200
then Bell would provide his experience and expertise,
commission the requisite ships and run the service and make
everyone concerned handsome profits. It does not seem that his
offer was taken up.

It must be said however that through communication
between the East and West coasts had already been greatly

facilitated by the coming of the *Comet* and her rivals. The Forth and Clyde Canal Company placed an advertisement in the Glasgow press in June 1813 arguing that, as steamboats now served Glasgow, Greenock and Gourock, passage boats plied between Glasgow and Paisley and as packets left the Clyde for all parts of Ireland and England, the canal was thus the best and most convenient route for those wishing to journey between Edinburgh and the west country. [35] In the following year, on at least one occasion, it was possible to sail by the power of steam all the way from Greenock to Edinburgh. The *Glasgow Courier* noted:

> A gentleman from Greenock on Thursday last [8th September 1814] made his journey by one steamboat to Grangemouth, and thence by another to Newhaven, a distance of 72 miles in eleven and half hours, for the trifling expense of 11 shillings and sixpence.[36]

When the *Comet* went into service between Grangemouth and Newhaven her schedule was designed to connect with the Canal Company's passage boats. With a morning sailing from Newhaven, a fishing port on the outskirts of Edinburgh, she arrived at Grangemouth in time to meet passengers from the incoming canal passage boat and sailed again for the capital in the afternoon. This proved, for a time, to be a popular and well-supported service and one which was welcomed by the Canal Company as a useful adjunct to their operations. It provided an attractive alternative route into the capital to the established stagecoach services which connected with the canal boats at Lock 16 near Falkirk.

However, once again, the unfortunate Bell was to find that where he had pioneered others followed, and more importantly, followed with larger and better vessels which understandably enough attracted away his customers. By 1818 the Grange-mouth—Edinburgh service was being hotly contested by rivals. First of these was the paddle steamer *Tug*, a 73 foot long, 100 ton product of John Wood's yard at Port Glasgow, and with two 16 horse power engines an obviously potent rival to the much smaller and less powerful *Comet*. One indication of the power and capacity of the *Tug* is that she, unlike *Comet*, made her way to the Forth via the North of Scotland and the notoriously rough and treacherous passage through the Pentland Firth between the mainland and Orkney. The *Annual Register* for 1817 noted in its entry for 22nd October:

> An important application of steam-vessels has lately been made in Scotland, and it is said with the most complete success . . . a Company in Leith have equipped a powerful steam-vessel, or tracker, possessing extraordinary strength, and completely adapted for encountering stormy weather. This vessel, which is most appropriately named the Tug, is meant to track ten other vessels, alternately, which have been peculiarly constructed by the same company, for carrying goods along the canal.
>
> The Tug, which may thus be compared to a team of horses in the water, tracks these vessels between Leith and Grangemouth, the entrance of the canal, along which they are tracked by horses. But the utility of the Tug is not confined to tracking; she has also two commodious cabins, and from combining the two purposes of tracking and conveyance of passengers, she is able to convey the latter with a degree of cheapness. . . the passage in the best cabin being for a distance of 26 miles, two shillings; and in the second, one shilling.[37]

Tug clearly offered economies by combing the towing function for canal cargo boats with a passenger service and it will be noted how prices for steamboat travel were falling, with a twenty-six mile journey by *Tug* costing as little as one shilling. *Tug* was not only a highly successful ship but she also gave her name to every later vessel designed to tow ships. The first recorded use of the word "tug" to indicate a towing or tracking vessel is in fact the extract from the *Annual Register* quoted above.

The pressures of competition were growing on Bell's ship. In the summer of 1818 the *Glasgow Herald* carried an advertisement for the *Grangemouth* steamboat offering "cheap and pleasant conveyance" with a daily departure from Newhaven at 8am, returning from Grangemouth at 2pm. This service landed passengers, by boat, at North Queensferry in Fife as well as at Bo'ness, thus providing a useful cross-river link. Using the *Grangemouth*, and travelling steerage both on her and on the Canal Passage Boats it was possible to travel from Glasgow to Edinburgh in one day for only four shillings. Such competition was too much for Bell and he withdrew *Comet* from the Firth of Forth and looked for other areas where opportunities for profitable trade might be more easily found.

CHAPTER 8
The Hebridean &
Highland Steamship

Within a few years of the *Comet*'s first trip in August 1812 Bell and his rival steamboat owners on the Clyde and the Forth and, indeed, contemporary operators on other major rivers such as the Thames, Mersey and Humber were providing an increasingly important passenger and cargo service. The steamboat, though so quickly successful, supplemented rather than supplanted the sailing ship. For many decades the sailing vessel would still form a highly significant part of the nation's transport infrastructure.

Comet and the other early steamers were river or, at best, estuarine traders and the deep sea and long-distance coastwise trade was not attempted. The sailing vessel was still the only realistic choice for such routes and this was particularly true where the carriage of bulky goods such as coal, grain, building materials, timber, etc was concerned. In this field the sailing vessel was to reign supreme for many years. In Scotland cargo carrying gabbarts and smacks would pick their way among the islands and sea-lochs of the West Highlands for the rest of the century, only gradually being replaced from the 1880s onwards by the steam puffer.

There was however, from the earliest period of steam navigation, a demand for faster and more reliable services for the carriage of less bulky and higher value goods, in particular the necessities of everyday life, perishable goods and of course the mails.

Despite this latent demand for an express service there was, in the early years of the development of steamer services in the West of Scotland, very little temptation for Bell or his rivals to venture beyond the Firth of Clyde. There was an obvious and profitable market in simply running these new services up and down the river between Glasgow and Greenock. On such routes the steamboats' advantages of economy, speed, reliability and comfort swiftly ensured their commercial success and they quickly replaced the fly-boats and stage coaches. There was thus, at first, little financial motivation for owners of the new steamboats to look for other markets. In any case the new method of propulsion had to be tested and had to win the confidence of the public before it was likely to be put into the more demanding deep sea service. In addition it had to be acknowledged that the ability of ships like the small and underpowered *Comet* to safely and usefully trade much further afield was highly questionable.

Soon however, the more enterprising owners sought to extend their horizons. In May 1815 an advertisement appeared in the *Glasgow Courier* offering what promised to be a quite novel, indeed a unique, experience:

> Marine Excursion from the Clyde to the Thames.
> A select party, not exceeding six, may be accommodated in the *Thames* schooner, late the *Argyll* steam engine packet. The vessel has received many improvements, is perfectly sea worthy, and enabled to proceed either by steam or sails, separately or united.
> Those who wish to enjoy this novel and interesting trip along the Coast of Scotland, Wales and England, will please apply to William Ker, the Agent, Broomielaw. The vessel will start early next week.[1]

James Cleland's list of early steamers[2] shows this ship as being a 72 footer launched from the Port Glasgow yard of Alexander

Martin & Co in April 1814 and equipped with a 14 horse power engine by James Cook of Tradeston. It is the second vessel on his list of Clyde steamers to have left Scottish waters and to make the passage to London. The first was the *Margery* which went south, via the Forth and Clyde Canal and the East Coast in November 1814, and traded on the Thames for a time before being sold to French owners for service on the Seine under the name of *Elise*, arriving in Paris in March 1816.

Like the *Margery*, the *Argyll* or *Thames*, as her name change suggests, was being transferred south to trade on the River Thames as part of the dispersion of steam power technology and expertise from its first British home on the Clyde. In the *Argyll's* case her owners obviously felt confident enough about the ship's reliability and safety to feel able to advertise for fare-paying passengers for the southward journey. The *Argyll's* West coast route, which involved a potentially difficult passage round Lands' End exposed to the full force of the Atlantic Ocean, was a more demanding proposition than the East coast route. It was however a more attractive voyage for the tourist and apart from providing excellent opportunities to view the picturesque beauties of the Scottish, Irish, Welsh and English coasts it surely would reward the courageous traveller with material to dine out on for many years.

It could be taken from his comments that Henry Bell had some intcrest in one or other of these vessels. It is perhaps more probable that he may have acted as an agent in their sale. In a letter published in 1825 he said:

> I sold the first steam boat that went to London, and also the
> first that went to France.[3]

This statement is obviously ambiguous and it is not clear from it if he was claiming to be the owner or just the intermediary for the sale and of course the phrase "and also the first" could be interpreted as referring to two separate ships but perhaps most obviously would be interpreted as speaking of one vessel, namely the *Margery*. However from other evidence it would appear to have been the *Argyll/Thames* that he was concerned with. In a letter written in 1819 to Sir Hugh Innes he says:

> . . . I have bin the first person that opened up all the Diffrent Rivers in Scotland and sent the first steam boat to London which is called the thames—I sold her to a Mr Dods and company she did well for the first year she payd her self and all expencces—she cost 3500 including the expences of taking her round the West Coast to London.[4]

The naming of the ship, and of Mr Dodd who was to command her on the passage south and the reference to the West Coast route all clearly indicate that Bell was involved with the *Argyll/Thames*, the second ship to go to the River Thames, rather than the *Margery*, the first. However the question of the French connection is still obscure as there is no indication that *Argyll/Thames* ever transferred to France. It may well be that Bell had an interest in both *Margery* and *Argyll/Thames* or was simply confused with which was the first steamship to sail south to the Thames or was, not untypically, making a good story a little better.

By good fortune the voyage of the *Thames* is unusually well documented and affords an interesting insight into the conditions of long distance steamer travel at this time. The House of Commons Committee on the Holyhead Roads—charged with investigating all aspects of the mail service to Ireland—took evidence from Captain Dodd of the *Thames* who reported:

> I beg leave to inform the Committee that I have just arrived in the port of London from the port of Glasgow, in Scotland, and that I have performed the voyage round the Land's End in a vessel that is propelled either by steam or sails, separate or united, and have experienced some extremely heavy gales of wind and high seas, and I found her more sea-worthy than any vessel I ever was in; she is fully capable of going head to wind in violent gales, and over high seas. . . This

The Thames *steamer, in a gale off Port Patrick on her epic
voyage from the Clyde to the River Thames
(National Library of Scotland)*

voyage has demonstrated that steam-engines are applicable to propel vessels at sea, in all kinds of weather.

Dodd went on to tell how *Thames* had outperformed all the mail packets and other sailing vessels that she had met on her 1500 mile passage. He went on to say that she was now sailing between London and Margate and on that route was regularly outrunning the Margate packets. In answer to a question from the Committee about the engine noise Dodd asserted that:

> There is no inconvenience from the noise; it does not make so much noise as a gentleman's carriage.[5]

This same pioneering voyage was the subject of an article by J C Delametherie which first appeared in the French *Journal de Physique* and was reprinted in the March 1816 issue of the *Scots Magazine*. The article describes the complement of the vessel thus:

> The command of the vessel has been given to Mr G Dodd, a young man of great resolution, who had gone to Glasgow expressly to bring it to London. He had made his apprenticeship in the English navy, and had afterwards distinguished himself as a civil engineer, an architect, and as a topographer. His equipment consisted merely of a master, four sailors of the first class, a smith, a fireman and a cabin-boy.

The startling impression that was made on both sailors and the general public by this new means of transport is graphically recounted by Delametherie:

> We arrived at Plymouth on Tuesday the 7th June. We were an object of astonishment to the sailors, who all collected on the sides of their vessels, to gaze upon us, and make their observations. We had no sails; our wheels were invisible; and as the fire happened at the moment to burn without

smoke, it was certainly difficult to divine the cause of our rapid motion.

From Plymouth we sailed without interruption to Portsmouth, where we arrived on Friday 9th June, at nine o'clock in the morning, having made 150 miles in 23 hours. At Portsmouth, the admiration was still more marked than elsewhere. The spectators crowded by tens of thousands, and the number of craft that pressed around us, became so considerable and inconvenient, that we were obliged to apply to the Admiral for a guard to maintain the police.[6]

The bold decision by the owners of the *Thames* to sail south and take fare-paying passengers is surely a very significant indication of the growing public and professional confidence in the new method of propulsion. A further expresssion of this confidence is seen in the ever wider range of excursion sailings being offered by the Clyde steamers. In 1812 and 1813 sailings went down river only as far as Greenock, Gourock and Helensburgh. 1814 saw the extension of steamer services into the wider waters of the Firth with locations such as Largs on the Ayrshire coast being served. However by 1815 one notes the advent of regular sailings to Rothesay, on the Island of Bute, and the start of services to the outer Firth and to the sea loch towns of Campbeltown, Lochgilphead and Inveraray. On one of these excursions the *Greenock* steamer carried out a pleasure sailing round Ailsa Craig at the request of some inhabitants of Campbeltown.

Towards the end of the 1815 season the *Greenock's* owners advertised a forthcoming sailing to Inveraray pointing out that:

. . . the scenery of the Clyde, the Kyles of Bute and Lochfine, are excelled by none in Europe. It is particularly grand at this time of the year, and the beauties and grandeurs of Inveraray Castle, the Pleasure Grounds and Deer Park exceed description, and all respectable visitors are allowed to see the whole. Refreshments can be obtained on board, and accommodation at Inveraray for Three Hundred People.[7]

Such pleasure sailings by passenger steamers, however important they might be in opening up the area to a new and important tourist trade, were no substitute for regular freight services, and they obviously did not as yet venture beyond the Clyde estuary.

Bell's perennial problems with better financed competition had forced him to look for other routes and this combined with his own enthusiastic vision of the role and the future of the steamboat must have inclined him to try to expand his sphere of operations into an area as yet unexploited by the new mode of transport.

The West Highlands would have been a very obvious field for such an extension of steamer services. The West Coast and the Hebrides were areas whose natural links had always been by sea rather than by land. From the time of the settlement of Dalriada by the Scots from Ireland, through the period of the Norse overlordship and the semi-independent Lordship of the Isles the West Coast communities had been bound together by the sea. Access to these waters from the Clyde had, until 1801, involved the long, potentially dangerous and frequently stormy passage round the Mull of Kintyre. The journey from Greenock to Fort William by this route involved a passage of about one hundred and ninety miles. However in 1801 the nine mile long Crinan Canal across Knapdale in Argyll had been opened and afforded a passage from Loch Gilp, an arm of Loch Fyne, to the Sound of Jura and cut a very significant seventy miles, to say nothing of the most dangerous part of the journey, off the passage. The Canal was to become increasingly important and in 1819, the year Bell set *Comet* plying to Fort William, there was recorded a total of 2028 passages through the waterway—almost double the trade recorded three years earlier.[8]

William Thomson, the resident engineer of the Canal Company at Crinan, had known Bell at the time of the building of *Comet* and had corresponded with him afterwards on the question of a Highland steamboat service. He wrote to Edward Morris and gave him the following account of Bell's plans for a West Highland service:

It was in 1818 our acquaintance was again renewed; and a correspondence took place regarding the extension of steam-communication through the Crinan canal, to the west Highlands, a favourite scheme of Mr Bell's. Steam-intercourse had by that time been established between Glasgow and Inveraray; and seeing it would be beneficial to the Highlands, and remunerating to whoever took up the trade, I entered into correspondence with Mr Bell on this subject, which he the readier entered into, that he and his *Comet* were driven from the Clyde by superior vessels under companies and capitalists, with which he could not compete.[9]

Undoubtedly feeling that, at a mere 42 foot in length, *Comet* was not really suited for the much rougher and more demanding conditions that would face her in service in West Highland waters, Bell had his ship considerably lengthened in 1819. Her new registered length is given as 73 feet 10 inches [10]. This was still well within the 88 foot maximum length restriction on ships using the Crinan Canal. This ambitious, or possibly over-ambitious, rebuilding was carried out to Bell's instructions by James Nicol on the beach at Helensburgh.

William Thomson, with what we may feel to be a practical engineer's reservations about the sea-worthiness of the *Comet*, sums up the effect of this work:

Mr Bell made his appearance with his *Comet* under the best repair she was capable of, and with such improvements as the engine and machinery would admit of, in August 1819, and proceeded to Fort William, between which and Glasgow he maintained a communication till late in that year.[11]

Bell advertised his enterprising new service under the heading of:

Cheap Conveyance to Fort William By the *Comet* Steam Boat
The *Comet* is appointed to sail from Glasgow to Greenock, Gourock, Rothesay, Tarbet, Loch Gilp, Crinan, Easdale, Oban,

171

> Port Appin and Fort William, on Thursday first, Sept.2, 1819, for the above places, at 9 o'clock morning, and to continue during the season every Thursday from Glasgow and from Fort William on Monday.

The fare for the journey from Glasgow to Fort William was £1.2.0, cabin and £0.15.0, steerage.

In his advertisement for this service Bell stated:

> In commencing this undertaking, which will be of so much utility to the public, the proprietor intimates to the merchants and gentlemen of the north-west coast of Scotland, that he intends to dispose of a number of shares of said steam-boat, which he hopes will turn out to be advantageous to the owners, and to the public at large.[12]

It may be conjectured that Bell's desire to sell shares in *Comet* arose both from a need to raise funds—the lengthening and refitting of his ship must have been an expensive project—and also from a wish to be able to invest in further shipping projects.

In late 1819 and early 1820 Bell was actively engaged in making a successful start on the Fort William service. This involved his selling off shares in *Comet* and floating the Comet Steam Boat Company. He was also engaged in discussions and correspondence about a far more ambitious project—the provision of a steamboat to serve the Hebrides, and specifically the port of Stornoway on Lewis. This Hebridean steamer project involved both J A Stewart Mackenzie of Seaforth, one of the main landowners on the island of Lewis, and other prominent West Highland landowners such as Sir Hugh Innes of Lochalsh in Wester Ross.

Quite how ambitious an undertaking Bell's Hebridean steamboat plan was is shown by a lengthy article published in the Edinburgh newspaper *The Scotsman* in 1822. The immediate occasion for the writing of this article had been the publication of the report of the House of Commons Select Committee to which earlier reference was made. The writer observed that:

> Hitherto it has been supposed that the advantages of steam navigation must be solely or chiefly confined to rivers, bays, and sea-coasts, and that it could be employed only in calm weather, and for short voyages. But the publication of the report opens up to us much more extensive views of its utility. It shows that its advantages are almost equally great on the main ocean as on a river; and that when perfected by future improvements, there are probably few voyages which will not be within its range.[13]

It is noteworthy that three years before this article was written Henry Bell was planning to send a steamship out to Stornoway, a passage of around 180 miles from Fort William through the notoriously stormy waters of the Minch to an island group lying 40 miles out in the Atlantic Ocean from the Scottish mainland.

The Hebridean steamship idea attracted the attention of a wide range of interested parties, including William Thomson of Crinan, who wrote in June 1820 to advise Seaforth on the size of ship that could pass the Crinan Canal. Thomson took the opportunity of pointing out that if a ship were to be large enough to sail with safety out through the Minch to Stornoway then it would be too deep in draught to sail up-river to Glasgow and recommended that any such service should be based on Greenock as the Clyde terminus.

A letter from Bell to Sir Hugh Innes of Lochalsh, dated 23rd December 1819, which survives among the Seaforth estate papers, gives a very valuable and clear account of Bell's activities as regards the *Comet* steamer service to Fort William and his proposals for the Hebridean steamship project. Clear, that is, if allowance is made for Bell's spelling and grammar, to say nothing of his breathless style:

> Sir
>   According to my promise I should have wrote you long

or this—but I hope you will excuse me when I give my rassons for not performing my promise—I was taken badlay and the natur of my truble was in pairt watter and a groath which I had to be cutt for and since I am gott better—Mr Crawford called on me today to be informed if my *Comet* steamboat was still going to Fort William—she only came in to the broomlaw [Broomielaw] on Wednesday last bein the last trip for this sasson as their was not passangers going to pay her expenses—and as this is a sasson that in general is very stormy weather—and I am going to give her a new sett of boylars—as the publick tack in to their head that those I had was not safe so as to daw away with that pradigest [prejudice?] I have gott new one althaw the old one is the best for standing the salt water. . .[14]

and so on for several pages.

Bell makes the point to Innes that he has sold shares in *Comet* and that it will be easier for a company than for an individual to keep her going even if, as might be the case, in the unpromising early months of the year she does not return a profit. He claims that since putting her into service on the Fort William run she had covered all her expenses and had made a profit of 10% on her valuation, which he puts at £1800 at least.

He goes on to outline his proposals for a boat to serve Skye and Lewis. He rejects the idea of changing boats at Crinan because of the difficulties of shifting luggage from one boat to another and the possibility of the connecting vessel not being there on time and passengers being held up. Coming to the financial point he writes:

. . . if you and any of the west hillans gentlemen wold join me I could put a steam boat to the Iland of Skay and Lewes by the first of April if not shooner. She is a good steam boat a will hold upwards of 80 to 90 pepale below Decks. She is a Main Cabin and steerage two Dining Cabins Stuarts [steward's] room with good water closets—I am just now putting in two good new copper boylars in to her and her Ingen is new allsaw—I vallow her at 2500 pounds stg. so if you and the reast of the county gentlemen wold take 1500 pounds in to her I would reatain 1000 in my own hands so as it might convence you that I not only had my contrays intrest at hart but that I am sertent it will pay well. . .

He points out that this way would be the quickest that the mail could be carried and that in his view another steamboat would be soon needed—which should be built to carry at least 30 to 40 tons of goods besides passengers. However he recommends the plan of starting with one boat because, as he points out, once the landed gentlemen of the area see the advantages to be brought to their districts by this first boat:

> . . . in bringen as it wer those Ilands within two or three days journee of Glasgow. . .

there would be no problems in raising funds for the second boat. Bell goes on to state:

> The advantages arising from out this conveyance is too many for me to Describe in this letter but what a vast Diffrance mead on to Fort William—by the small steam boat called the *Comet*—I hop it will by this time convince the County Gentlemen in that pairt of the countray the grate advantage is to derive by that moad of convayance beside she hast allways ceap her time of sailling and never missed one day which no other steam boat in the river Clyde can tell the same storrie—for I flatter my self that I am upon the bests and simplests prinspal of any. . .

He ends with a summary of the financial position:

> . . . if my view meats with your idays I wold be glad to hear from you—as mony is object to you it might imbark into it yourself—I wld only requair £1000 in course of a month and I wold lett the other 500 remain till Lammas so as the boat wold have ganed near as much.

The last sentence was an offer to defer payment of the balance of £500 until the Lammas quarter day (1st August) when, he suggests with his usual confidence, the new ship would have earned her owners almost as much as that in profit.

Bell must have intended to place an existing ship in which he had an interest, or one which he could be sure of purchasing quickly, on to this new route. In no other way could so speedy a start be made to the Hebridean service. Indeed in a postscript to his letter he tells Innes that it would require six months to build a new boat and that he would wish such a new boat only to be built during the dryer weather of the summer on the grounds that a steamboat requires well seasoned and dry timber. He went on to explain that the steam escaping from the ship's engines and boiler was likely to cause rot if the timbers are not properly seasoned and dry.

It is not known what vessel Bell proposed to use for the island service—certainly it could not be the *Comet*. Whatever obscurities there are in the letter he clearly distinguishes between the *Comet*—which he valued at £1800 and the other vessel valued at £2500. The fact that in both cases he refers to new boilers seems to be simply coincidence. His reference to the benefits brought to Fort William by "the small steamboat called the *Comet*" would seem to distinguish beyond any doubt between the proposed boat for the service to Lewis and the *Comet*.

Innes and the West Highland gentlemen obviously did not immediately fall in with Bell's suggestions. The matter of the Hebridean steamship was to drag on through 1820. Innes wrote to Seaforth in May 1820 reporting that:

> I have been corresponding about steam boats. There is a difficulty about getting one of a proper size at present for a passage to the Lewes.[15]

and that he understood that:

> ... Mr Bell is now at Fort Augustus or Inverness making arrangements for a boat from Inverness to Fort Augustus—

and a carriage from there to Fort William.

He goes on to comment that such a service, which as we shall see was to be operated by the *Stirling* steamer, and which was intended to connect with the Glasgow to Fort William steamer service being provided by *Comet*, would be a great accommodation to that part of the country:

but to Stornoway nothing.

This situation clearly caused Innes to look into the matter on his own account—later in 1820 the *Glasgow Chronicle* was to report:

We understand that Sir Hugh Innes Bt., has made arrangements for plying a large steam boat, calculated for the conveyance of bulky commodities between Kyleakin in Lochalsh, and Glasgow, once a week to commence early in Spring; and that it is in contemplation to join another boat, which will ply between Kyle and Lewis, taking in its course the various inlets in Skye, and on the mainland.[16]

Before this decision was taken further correspondence had passed between Bell, Innes and Seaforth during 1820. Seaforth wrote to Bell in July from Seaforth Lodge on Lewis:

Sir,
   Mr Anderson of the Bank [in Inverness] informed me before I left Brahan [Brahan Castle, near Dingwall in Easter Ross, the main seat of the MacKenzie's of Seaforth] that you were afraid you would find yourself under the necessity of giving up the Loch Ness Steam Boat from want of sharers in the concern. I should be sorry your enterprise should be thus early damped.
   It has however for sometime been a very great object with me to see a similar attempt, made between Glasgow and Stornoway. . .
   My object in writing to you at present is to induce you to

177

immediately come to Stornoway and visit these coasts with a view afterwards to commence this Speculation, which I am <u>quite satisfied</u> under proper management could not fail to pay. . .

Seaforth painted a picture of the large tracts of land which would be opened up to commerce by steam navigation and went on to give Bell his views on the type and frequency of ship and service that would be needed. Seaforth, perhaps with the benefit of Thomson's advice, was inclined to favour the idea of a service running from Crinan to Stornoway and using Crinan as a transhipment point for goods to and from Glasgow. He points out the need to involve the merchants and shopkeepers of the islands and West Coast and ends:

> With this brief outline I shall close, adding that supposing the Capital to be vested in the concern were to be £3000 I should consider your holding 1/3rd as indispensible and that a committee of arrangement should after each trip credit the Expense recd. and disbursements.[17]

Bell replied on 16th August 1820 to Seaforth. Bell acknowledges Seaforth's letter and commended his wish to bring a steamboat into service to the Western Isles:

> . . . I sertnlay think it wold be one of the best improvements ever ceam in to that pairt of the countray. . .[18]

and goes on to bemoan the condition of the people of the Islands who have no wish to improve their condition.

With what might be described as a certain degree of courage, or indeed recklessness, he castigates the landed interests of the Highlands and Islands for their previous inaction in such matters and for their wasteful tastes and preoccupations:

> . . . you landed Gentlemen ought to dow a grate deal more

178

than you dow in forming improvements in your Ilands and coast of the Highlands... the most pairt of the land gentlemen is so much taken up with politicks, gamblin and other trifling amusements that they both neglect their own Intrest and the Intrest of their countray...

This may not have been the most obvious or tactful line to take to a great Highland landowner but it is indeed a judgement that many modern historians of the Highlands in the years after the Jacobite Risings would not find much argument with.

There had been many radical changes in the Highlands after the '45, changes which were accelerated by the ever-growing effects of the Clearance movement. Although both of these pressures had irrevocably altered and undermined the traditional relationship between the clan chiefs and landowners and their followers or tenants, traditional values still were evidently of considerable significance in this part of the world. Bell, it would seem, viewed the support of key opinion formers, such as the clan chiefs and other landed proprietors, as an essential pre-requisite to the success of any Highland project such as the Hebridean steamship. He pointed out to Seaforth that most of the Highlanders would:

... depend whole on the Judgement of a few pipale in genral bred up to the land...

and enthusiastically wrote:

... what a grate Diffrance wold be if each landowner war coming forward with marchants, stor ceapers [store-keepers] and fishermen in small improvements for the good of the Countray—it wold go forward with a quite Diffrant spirit for they wold be saying our laird is taking an intrest in this...

He went on to point out to Seaforth the futility of outside agencies

179

attempting to do anything in the Highlands without the support of local people and ensuring the engagement of the general interest of the district—a point of view which would be more than echoed by other would-be Highland improvers both before Bell's time and since his death.

Bell's remarks were perhaps better targeted and less risky than might at first seem the case. MacKenzie of Seaforth was, despite his name and territorial title no more a Highlander than Bell himself. He was the second husband of the eldest daughter of the last Lord Seaforth and came from a family of Galloway gentry, the Stewarts of Glasserton. He might thus have been not unsympathetic to Bell's very typically Lowland comments about the backwardness of the Highlander and the apathy of the other Highland landowners.

What Bell had in mind to:

> . . . obtain the grate object of cheap and expedishous conveyance of cattle pipale and goods between the western Ilands and the low countray. . .

was a steamboat which would ply to the islands once a week for eight months a year, and twice a month during the four winter months. Such a service was, as he explained, badly needed as the existing commerce of the area was very backward and Highland merchants were, for want of fast and regular communications, unfortunately obliged to either lay in costly and excessive amounts of goods or face the risk of being out of stock for months. The effect of this deficiency in transport arrangements was to raise prices for all the necessities of life.

Bell pointed out to Seaforth that through his actions he had opened up steam navigation in Scotland and other countries and that he would have no objection to engage in such a venture as the Hebridean steamship. He made a point of insisting, in a typically colourful passage, that the local landed gentry in the areas where the steamboat would trade would have to take shares in the venture before he would consent to invest in it:

180

> ... without the countray gentlemen wold take an intrest I
> wold have nothing to dow with it for the highland pipale is
> all very clanish to one another and if their chiefs had no intrest
> in it they wold to a sertentlay rather walk 40 miles than pay
> one shilling—and by this old custom in a grate degree the
> Improvements of the highlands and Ilands is so far behind. . .

Bell's view of what was required for this service was somewhat
different from Seaforth's. MacKenzie had in effect envisaged a
branch boat running from the existing Glasgow—Crinan service.
Bell, as always, takes the broader view of the situation and argued
the more ambitious case for the construction or purchase of two
ships, one of 160 tons burden:

> ... a proper sea boat ... with a sufficience of strength of an
> ingen to take her through the watter ...

this would cost some £5500, and secondly a smaller vessel of:

> ... about 80 tons burden that wold pass throw the Crinan
> Canal by Lochgilphead to Glasgow direct ...

which was estimated to cost around £3500. Presumably the larger
ship would be based on Fort William and serve Skye and the
outer Islands with the smaller ship completing the link between
Fort William and the Clyde. In view of Sir Hugh Innes's
involvement in this project it may also be reasonable to think
that these ships would ply to his West coast estate at Balmacara
in Wester Ross on the shores of the Kyle of Lochalsh between the
mainland and Skye.

Bell suggested that Seaforth should give him:

> ... an order to go forward with either the first or largest one
> or the second—with liberty to me draw on you as the veshal
> goes on for such sums as shall be agreed upon by way of
> instalment. . .

Bell would meet one third of the cost of the boat until she was ready to go into service, at which time shares of £50 or £25 would be sold. With the obvious benefit of hard won experience and with typical shrewdness he points out:

> . . . if you wait till you gett a subscription paper filled up it may be a year for a grate deal of pipale daws not understand a thing till they can feel or see it—so it hast ever been my practis for it to mount the boat and then seek for partners. . .

Bell stated that, had it not been for the fact that much of his capital was tied up in other ships, he would not have hesitated to make the entire investment himself and confidently asserted that there was no risk to Seaforth or his friends in such an investment and went on to remind Seaforth of the success of his Fort William steamboat, the *Comet*.

He ended his letter by emphasising the great advantages that would accrue to merchants, fishermen and the whole community by the advent of such a regular steamboat communication between the Islands, West Highlands and the industrial areas of Central Scotland. As was characteristic of his tone throughout this letter Bell ends on a typically robust and assertive note:

> . . . if I had such property as you it wold not stand me a moment thought but cause it to be done. . .

Perhaps, somewhat belatedly, thinking that he might have been rather over-bold in his critical comments on the deficiencies of enterprise on the part of the Highland lairds, to say nothing of his accusations of their preferential attention to politics and gambling and such "trifling amusements", Bell wrote a brief note in the margin of the letter observing that what he had written:

> . . . is perhaps rude but I hope you will forgive the liberties I take.

182

Whether Seaforth took offence or not is not known but Bell's approach had little of the deferential in it and he clearly had no inhibitions about treating with the landed gentry on terms of equality. In a period of considerable social formality which laid great emphasis on social status this approach is perhaps indicative of Bell's innate self-confidence and his frequently expressed self-image as the pioneer who had performed great deeds for his country.

Despite all Bell's enthusiastic advocacy the project did not take place quite as planned, or indeed with his involvement. In the Seaforth papers for April 1821[19] there appears a list of subscribers to a Hebridean steamer—a list which did not include Bell, but which did include such noted Highland grandees as Lord MacDonald, MacLeod of MacLeod, as well as Seaforth and Hugh Innes of Lochalsh. Their planned vessel is presumably the ship to which the *Glasgow Chronicle* article of November 1820 quoted above refers. Even with the support of such important and influential subscribers there was evidently a delay in getting the service started, for in June 1821 Sir Hugh Innes wrote to Seaforth:

> I am sorry you have made no arrangements . . . about a steam boat plying this year to Stornoway. . ?[20]

and went on to speculate on the possibility of some of the other ships which were already in service in the West Coast area being able or willing to extend their operations to the Islands and to the North West Highlands. The reason why Stewart MacKenzie and his associates did not take up Bell's offer, which would seem to have assured them an earlier start to the service than their own alternative could, is not recorded.

While these long-running and complex discussions and schemings were being carried forward we must not forget that Bell was still involved with the operation of the Baths Inn and also had the responsibility for the more everyday work of running a shipping company.

Bell took full advantage of every suitable opportunity that presented itself to promote his steamer service and find new markets. His rival John Thomson had criticised Bell for what he saw as his vulgar promotion of the *Comet* in Edinburgh in 1813. Despite such strictures Bell obviously held to his faith in the power of advertising and saw the potential income which could be had from running special events and services. In June 1820 he arranged for the charter of the steamboat *Fingal* to supplement the *Comet* and planned special sailings to capitalise on a major annual event at Fort William. He advertised his plans as follows:

> The Great Wool and Cattle Market at Fort-William
> The proprietor of the COMET STEAM BOAT, with a view to the accommodation of his Friends, and Merchants going to FORT-WILLIAM WOOL and CATTLE MARKET, which begins on the 14th June current, has engaged the FINGAL STEAM BOAT, Captain Dick, to sail from the Broomielaw on MONDAY NEXT, at Five o'clock Morning, for LOCH–GILPHEAD, where she meets the COMET, which will carry the passengers round to FORT-WILLIAM, and land them on Tuesday Afternoon. . [21]

The *Comet* spent a busy week. On the Monday, when she would normally have sailed from Fort William for Glasgow she only went to Crinan, returning the next day to Fort William. On Thursday 15th she again sailed south for Crinan, with a disappointing total of only four passengers but on the Friday brought 56 passengers from Crinan and intermediate stops back to Fort William and the wool and cattle market. On the Saturday a trip to and from Oban was made, presumably to return market visitors home to the Oban, Appin and Lismore area. The *Comet*'s normal service to the Clyde was resumed on Monday 19th June.

Bell clearly had a good eye for a business opportunity. However he, Innes and Seaforth, were by no means the only people to have an interest in the potentially rich market presented by the Highlands and Islands. At the end of 1820, under the heading "Steam-packets to the Hebrides and Northwest of

Scotland", the *Greenock Advertiser* noted that:

> The great facilities now afforded for visiting many parts of
> this country, by means of the cheap and safe conveyance of
> these vessels, continues to be everywhere on the increase.
> At present it must afford much satisfaction to all interested
> in the Northern parts of our Island, to learn, that a
> communication is now to be opened, by this admirable
> invention, to many parts of the Highlands, which were lately,
> and are yet, comparatively inaccessible by roads. It is now
> intended that a Steam-boat shall begin to ply from the Clyde
> to the Lewes, through the Crinan Canal and Sound of Mull—
> to call at Tobermory—from thence to the Sound of Skye—
> call at Isle-Ornsay, Lochalsh, Castlemoil, Portree and
> afterward go on to Stornoway. The Steam-boat *Highland
> Chieftain*, has already gone as far as the Sound of Skye on
> this route, for a trial, and performed the passage in the
> remarkably short space of 35 hours from Glasgow—a
> distance of 235 miles, notwithstanding she had to stem
> currents which run so violently in the Sounds of Skye and
> Mull. She returned in nearly the same time, and encountered,
> with great intrepidity, very severe weather. The track now
> proposed that this Steam-boat shall run, will be highly
> gratifying in the summer months, for an excursion.[22]

The *Comet*'s route was subject to considerable modification to
meet passengers' needs. Calls were made at a wide range of ports
along the way—for example the April 13th north-bound sailing
took two passengers from Glasgow to Renfrew, and the service
a week later had one passenger who only travelled to Erskine
and another who went the mile further to Bowling. On
subsequent sailings calls were made at other upper-Clyde
settlements such as Dunglass and Dumbarton. The same pattern
of request stops was followed in Highland waters with
passengers being regularly dropped off at islands and ports
which did not feature on the original itinerary.

By February 1820 Bell had, as he had indicated his intention
to do in his first advertisement for the *Comet* sailing to Fort

William, succeeded in selling shares in the vessel—in fact 20 of the 36 shares in the *Comet* had already gone to 23 investors.[23] These were mostly drawn from the Fort William area and they included the type of people whose support Bell had indicated to Seaforth as being a prerequisite for success in such a venture; local landowners such as Sir Ewen Cameron of Fassfern and Fort William merchants such as Donald McIntyre and Alex MacDonald. In addition to the majority West Highland interests there were a number of other investors, including Elizabeth Bell of Glasgow, presumably Henry's sister, and Robert Bain, "mariner at Helensburgh", the skipper of the *Comet*, as well as a block of shares held by Margaret Bell's relatives the Dykes of Burnhouse.

This loosely organised co-partnery transformed itself into a more properly constituted company, the Comet Steam Boat Company. At a public meeting held in Fort William on 30th September 1820, the shareholders elected a committee and office bearers with effect from 1st January 1821.[24] This first committee was typical of the investors in the *Comet* and comprised:

> Sir Ewen Cameron, of Fassfern, Baronet
> Colonel McLean of Ardgour
> James McAlpin, Esq, Corpach
> Captain Neil McLachlan, Castleton
> Duncan McIntyre, Junr, Merchant, Fort William
> Robert Flyter, Esq, Sheriff Substitute of Fort William, *Treasurer*
> Henry Bell, *Superintendant*

The constitution of the Company set down the rules for the Captain of the steamboat to observe in the handling of the company's money etc.

> That the Captain keep a regular Book, in which he shall insert the Number of Passengers, Luggage, etc. to and from each place, and a regular book for his Men's time and wages:— that he also insert in a book an account of all Orders and

Disbursements on account of the Boat and Machinery. . .

Remarkably enough, in view of the later wreck and loss of the *Comet* this "regular Book" has survived and gives a detailed account of the number of passengers, and in some cases the nature of the passengers, for all the *Comet*'s sailings from 31st March 1820 to her wrecking in December 1820. The book also gives a detailed record of expenditure on the ship, from crew wages, to coal, canal and harbour dues right down to the expenditure of ten pence on July 12th for an "Oil Cann". [25]

The Company's Articles of Association, with a proper concern for command structures instructed that:

> The Captain shall have full power to discharge any of the Men under his command for insubordination or misconduct: and that no Proprietor, while on board as a Passenger, shall interfere in any manner with the Management of the Boat, but may report to the Committee in writing, any impropriety in the Captain or Crew he may witness.[26]

The document went on to state that:

> It being understood that Mr. Bell, who is presently Ship-husband for the *Comet* is to continue as such, to the exclusion of the Committee now named, till the said First of January 1821, at which period he is hereby directed and taken bound to deliver the Books and Accounts containing his intro-missions to the present Committee, to enable them to enter on the administration of their duty.

A ship's husband was the owner's agent who attended to those items which fell outwith the scope of a ship's captain and officers. He would deal with commercial matters relating to trade and cargo and with the administration and maintenance of a vessel on behalf of her owners.

While the new company was in the process of formation

Bell was actively engaged in selling off his remaining shares in the *Comet*. We have already noted in Chapter 2 his correspond–ence with Dr Archibald Wright on the sale of a £50 share in the ship.[27] In a note on the registry document, dated 3rd October, Bell's disposal of the remaining 10/36th of *Comet* was noted. Other than Dr Wright these purchasers were all from the West Highland area of operation of the *Comet*. All these transactions, like Dr Wright's which took place in August 1820, would seem to have been completed before the establishment of the Comet Steam Boat Company. The Company's Treasurer, Sheriff Substitute Robert Flyter is, for example, one of those who formed part of the second wave of purchasers of shares.

According to Bell's own report to Seaforth the *Comet* had done well in her first year of trading to Fort William and the enthusiasm with which shares had been taken up by landowners and merchants in the West Highlands does indeed suggest that she had proved to be a profitable undertaking.

The evidence of the Captain's account book is of an increasingly successful venture. Obviously trade was brisker in the summer months and the 3rd August sailing from Glasgow to Fort William saw the highest income recorded, with £69.7.0 being taken from 155 passengers. In the period of just over six months covered by the account book fares totalling £1910.1.0 were collected. Luggage earned another £6.3.0 and the company's receipts from the ship's steward came to £72. The total recorded expenditure on the ship's running costs, wages and dues only amounted to £739.2.11. Obviously the difference was not all profit—the *Comet* was a diminishing asset as she aged and provision needed to be made for her replacement and perhaps to service any debt that there may have been on her. Nonetheless it seems perfectly clear that things were going well for Bell and his ship. Indeed in his second letter to Dr Wright he had enthusiastically written:

> I plainly see Doctor we must have another steam boat as the two will pay better than one. . .[28]

Sir

*I was favour'd with a letter from
Mr McIntosh Informing me that you
wished to have a shair in the Comet
Steam boat — which you shall have
as follows one shair is Just now £55 for
but you Deduct 20 persent from — a
fifty pound shair as usual is £10 for
which lives to be paid Just now £45 for
and you will have a right to an
equal Divedion of what the makes from
the first of August 1820 — this is
my terms and the said £45 to be
paid Just now and for further
of your shair I leff you to be Judge
on the Back, of the regester and allow
on the general vindishion of the
Comet steam boat and your rights
will be as good as the other partners
if this is agreable to your views
please send me the last —*

*Bath 16th Augt
1820.
To Dr
Arch Wright Esqr
at
FortWilliam*

*I am your Most
obedient*

Henry Bell

Letter from Bell to Dr Archibald Wright regarding the sale of a
share in the Comet
(Strathkelvin District Libraries)

By October 1820 Bell had sold all 36 shares in *Comet* at a nominal price of £50 each. This would place the value of the ship at £1800. However as we saw in the case of Dr Wright's investment Bell said that shares were valued at £55 but he offered Wright a £10 discount which would bring the cost of a share down to £45. If we assume that similar prices were applied to the other investors the minimum income Bell would have derived from his sale must have been £1620.

The smaller of the two ships which Bell had proposed to Seaforth that should be built for the planned Hebridean service— the one of 80 tons burden and costing £3500—would have been substantially larger than *Comet*, which even after her 1819 lengthening only had a registered tonnage of just under 30 tons. While this rather esoteric figure of registered tonnage cannot be exactly compared to the 80 tons burden spoken of by Bell it does suggest that the value of the *Comet* was quite substantial for a small, rather under-powered, and certainly no longer new vessel. The money Bell derived from the sale of the *Comet* shares would certainly have been adequate to provide him with funds enough to take his part in the speculation with Seaforth, Innes and their neighbours in the Hebridean steamship venture, an enterprise in which a total cost of £9000 was envisaged. It was undoubtedly the assurance of the income from the sale which enabled him to write so confidently to Seaforth and to guarantee his share of the investment.

One must, however, enter the cautionary note that it is somewhat unlikely that the £1620 was all going to be available to Bell for such a project. It would be quite untypical of Bell if he had not had, at this time, his usual burden of debt and, although there is no surviving evidence either way, it is not unreasonable to speculate that one reason for selling off all his interest in *Comet* was to satisfy creditors as well as to fund other enterprises. It will be recalled that John Robertson said that the reason that *Comet* had to undergo what was virtually a "do-it-yourself" rebuild on the Helensburgh beach was that Bell still owed John Wood money from 1812, and of course the debts to Robertson

190

himself and to David Napier were still outstanding.

The sale of the shares did not mean however that Bell and the *Comet* were no longer connected. Bell would appear to have been, even if he was no longer a part-owner of *Comet* and as such a partner and shareholder in the Company after 3rd October 1820, still acting as the Company's superintendent and the mainspring of the enterprise.

However disaster struck on 15th December 1820, even before the new administration had taken control. *Comet* had sailed from Glasgow to Fort William on Monday 4th December on her restricted winter service of one round trip a fortnight. Only 41 passengers travelled northwards on her. She would have normally been due to return from Fort William on Monday 11th December. However it is reported that she had a collision with a half submerged rock on the voyage north and a slight leak resulted. It is also reported that on the return trip south from Fort William she struck a rock off Sallachan in Loch Linnhe and was beached to allow repairs to be made. Sallachan Point is two miles below the Corran Narrows and about ten miles from Fort William.

Whether or not there were two accidents is not clear. The Captain's account book has an entry for December 8th "7 men, 1 day at Salouhan" and for the 10th there are entries for the hire of a boat from Fort William to Corpach, and for landing materials at Corran from the *Glenaladale*." On 12th December there is an entry for the hire of a total of 23 men for "docking, etc" and a similar entry for the hire of 10 men and the repair of a boat on 13th December.

The problem with these entries is obvious. The expenditure on 8th December at Sallachan presumably refers to the stranding incident. However the 8th seems very late for the *Comet* to be northbound in Loch Linnhe—having left the Clyde on the 4th—and is obviously too early for her scheduled southbound voyage, which was not due to start until Monday 11th. A possible explanation is that the report that *Comet* had sprung a leak on the northbound voyage is correct and that her normal

191

southbound service was cancelled and she was sailing for the Clyde for repairs when the stranding at Sallachan took place. According to press reports she was carrying passengers when she finally sank, but there is no record of passengers or fares collected on the final voyage in the Captain's account book. One person who was reported to have been on board *Comet* on her last voyage was Bell himself, who had apparently been in Fort William raising money for a new and larger ship.

After the work at Sallachan the voyage was resumed and an overnight stop was made at Oban where *Comet* arrived on the afternoon of the 14th December. The *Comet* at this point was taking in water and Captain Bain's cash book shows an entry for ten shillings expenditure for four men for one night's pumping of the vessel. The next day, 15th December, she pressed on southwards down the Firth of Lorne in a violent snowstorm. When *Comet* was coming into the Sound of Jura with the intention of making for the Crinan Canal and was entering the area know in the Gaelic as the Dorus Mor, the great gate, she ran onto rocks at Craignish Point. The *Greenock Advertiser* reported the accident in the following terms.

> We are sorry to state, that the *Comet* steam-boat, on Friday last, on her way home from Fort-William to Glasgow, at half past four in the afternoon, while passing through Dorish-more, at the point of Craignish Rock, was struck with a strong gust of wind, which laid her on her beam ends; and in ten minutes, owing to the great current of tide and high seas and wind, was laid broadside on the rocks. Every exertion was made for the landing of passengers and men, which was safely accomplished. On Saturday morning she was a complete wreck—It is understood that the Comet Steam-boat Company has contracted for another Steam Passage-boat, of greater power, to ply between Fort-William and Glasgow, which is expected to be ready by the 1st of April.[29]

The ship is reported to have split at the point where she had been lengthened in the previous year. Some sources have

Bro forward	697	12 5
Dec 8 1 Do 7 men 1 Day at Salachan	" 17	6
10 Boat from F William to Corran	" 3	"
Expence of Landing Materials at Corran from the Glenaladale	" 5	"
12 19 men 3 tides Docking &c	7 2	6
4 Do 2 tides 2/ ⅌ tide	1	" "
13 6 Do 8 tides	6	" "
4 Do 1 tide	" 10	"
for Mending a Boat Employed	" 5	"
15 4 men 1 night pumping at Dam	" 10	"
Pumpmaker Do	" 4	6
7 Barrels Coals Do	" 14	4
" 19 4 men 3 days at Craignish from Crinan with a Boat	1 10	"
21 Hire of a Boat from Craignish to Luing to look after wreck	" 5	"
whisky to Boatmen	" 2	"
22 Boat from Luing to Scarba & Craignish	" 7	6
whisky to Do	" 3	"
25 Two Large Boats 3 days trying to get up Engine &c	6 "	"
Paid Jas Campbell for hiring & Small Boat 5 days		
	£724	14 5

Comet *account book entries showing costs incurred in attempted salvage operations at Craignish.*
(Private collection)

suggested that Bell, plagued with his usual financial problems, had economised by using unsuitable timber for this lengthening and the total loss of the *Comet* has been attributed to this cause.

Strenuous attempts were made to salvage all that could be removed from *Comet*. Exactly how strenuous these efforts were is graphically shown by the series of entries in the Captain's cash book.

Dec 19th 4 men 3 days at Craignish from Crinan with a boat	£1.10.0
Dec 21st Hire of a boat from Craignish to Luing to look after wreck	£0.5.0
Whisky for boatmen	£0.2.0
Dec 22nd Boat from Luing to Scarba and Craignish	£0.7.6
Whisky to boatmen	£0.3.0
Dec 25th Two large boats 3 days trying to get up engine etc.	£6.0.0
Paid Jas Campbell for himself & small boat 5 days	£1.4.0
Paid Donald Brown for spars and six days	£1.0.0
Charles Mclean for five days	£0.13.6
Alex. Campbell for four days	£0.10.0
Dugald Campbell for four days	£0.10.0
Donald McLardy for 8 days	£1.0.0
4 men from Kirkton 3 tides	£1.10.0
3 men from Airdes & 3 Crinan	£2.5.0
Whisky from 15th to 25th December to people employed at Craignish	£5.18.6[30]

The hull could not be salved. The stern section drifted off soon after the accident, hence the hire of boats to go after it to the neighbouring islands of Luing and Scarba. The bow section remained on the rocks for just over ten days before slipping into deeper water, thus allowing some of the fittings to be saved. Craignish Point is quite an isolated area and any salvage operation mounted there would present considerable problems. The Craignish Peninsula has no large centres of population and even today the single-track road down the peninsula stops a mile or so short of the Point. As can been seen from these accounts workers and boats had to be hired from various locations. Kirkton and Airds are settlements on the Craignish Peninsula but other workers were ferried across from Crinan. Craignish Point lacks shelter of any sort and the seas around it are troubled by strong cross currents and a mid-winter salvage operation must have been a severe ordeal and one can well believe that the £5.18.6 worth, around 13 gallons, of whisky would be both needed and appreciated.

The story of *Comet*'s engine is an area of some confusion and controversy. Some writers claim that the original engine of the *Comet*, now in the Science Museum, London, was that which was fitted to her when she sank. Others follow John Robertson, the builder of the engine fitted to the *Comet* in 1812, who had a continuing involvement in the history of the engine, in agreeing that *Comet* had been re-engined. This theory holds that the Science Museum engine is the first engine and that a second engine was fitted, probably when she was lengthened in 1819. The idea of re-engining the *Comet* when lengthened seems quite logical. Donald MacLeod states that *Comet*'s engine was saved at Craignish and that he believed that parts of it were incorporated into the engine of her replacement the *Comet II*.[31]

This account, if true, would obviously conflict with the notion of the engine which is now in London being that from the wreck. It should be said that John Robertson reported that as far as he knew the second engine was not recovered from the wreck.[32] There is however no very obvious reason why Robertson should be particularly well informed about a second engine, an engine which he was not associated with, as opposed to the first engine which he had built. As we have seen prolonged attempts were made to salve the engine from the wreck and MacLeod's story that some parts were saved and re-used, which is doubtless derived from his uncle's personal knowledge of and friendship with Bell, may be well founded. With the hull split into two parts there could surely be no realistic prospect of doing more than recovering fittings from the hull and the only obvious target for the ten day salvage operation would be the engine. This would be a particularly desirable objective if the engine was indeed a replacement one and only just over a year old.

It will be recollected that when Bell wrote to Seaforth in December 1819 he had said that he was about to fit a new set of "boylars" to the *Comet*. William Thomson, the resident engineer at the Crinan Canal, who knew Bell well and wrote about him to Edward Morris agrees that part of the machinery of the *Comet* was recovered from the wreck[33], and Thomson, locally based, would undoubtedly have been in a good position to know exactly what had happened.

The Captain's account book is, for once, unhelpful. It shows expenditure on the hire of "two large boats, 3 days trying to get up engines etc" which is ambiguous but does perhaps tend to suggest an unsuccessful venture.

If MacLeod's account, endorsed by Thomson, is accepted then it is noteworthy that elements at least of both engines of the *Comet* have survived. The original engine was sold, as we are told in the version given to John Buchanan by John Robertson, to a Glasgow coachbuilding firm. Robertson said that he was later commissioned by a distillery firm in Greenock to purchase it for them at a cost of £60. Robertson reported that the engine then passed into the possession of the noted Glasgow engineering firm of Claud Girdwood, and was exhibited to a meeting of the British Association in Glasgow. It was certainly eventually to be presented by the Glasgow shipbuilding and engineering firm of Robert Napier & Sons to the Science Museum in London where it remains today. The original cylinder, which was replaced to provide additional power early on in *Comet*'s career, is now on display at the Scottish Maritime Museum at Irivine. The engine of *Comet II*, which may as MacLeod says, incorporate parts from the second engine of *Comet*, survived the loss of that ship and was displayed in Glasgow's Kelvingrove Park at the time of the *Comet* Centenary celebrations in the summer of 1912.

As the Greenock newspaper report stated, the Comet Steamboat Company was able, immediately after the wreck, to announce that they had another vessel of greater power on order and that this vessel was expected to come into service by 1st April. It would seem clear from the timescale, which would have left little time for design, tendering and construction, that the company had already planned this addition to their fleet even before *Comet* sank off Craignish.

William Thomson, writing to Edward Morris in 1843, recollected that Bell had indeed been a passenger on board the original *Comet* on her last voyage and had in fact:

> . . . accompanied her to make arrangements with subscribers about a new and more powerful boat for the ensuing season. . .[34]

The "new and more powerful boat" was to be built at James Lang's yard in Dumbarton in 1821 and registered at Fort William as the *Comet*[35]. The second use of the name *Comet*, while doubtless a sound commercial decision to capitalise on the name of the pioneering steamer and also a pleasant compliment to Bell, has nonetheless caused a considerable amount of confusion to writers ever since. Even contemporaries seemed to be easily confused— the writer of the article on steam navigation in the 1824 supplement to the *Encyclopedia Britannica* lists the new ship built in 1821 as if she were the original *Comet*, lengthened, rather than as a totally separate vessel.[36]

Comet II was certainly an improvement on the original. At 81 feet in length she was seven foot longer than *Comet I* as lengthened but was still small enough to navigate the locks of the Crinan Canal. Her 30 horse power engine would certainly have been a more effective instrument of propulsion than the much less powerful machinery fitted to her predecessor.

Henry Bell certainly had a close professional involvement with *Comet II* although her registry documents do not indicate that he owned any shares in her, and the exact nature of his interest in the boat is often said to be unclear. However there seems to be no real reason for any confusion when it is recollected that the ship was built for and owned by the Comet Steam Boat Company. Bell had completely sold out his interest in *Comet* during 1820 but nonetheless was listed in the Comet Steam Boat Company's papers as their Superintendent. Bell's name, his reputation as a pioneer and his entrepreneurial flair would have been of as much value to the Company as any financial investment he might have been able to make. There seems no reason to doubt that he continued in this capacity as Superin-tendent when the Company put *Comet II* into service and that it was his role as the public face of the Company which was instrumental in attracting support.

William Thomson wrote of this period after the sinking of *Comet*:

> A new boat was at this time determined on, towards which the gentlemen of Lochaber gave their warm support, by taking shares, not less from the advantages seen to accrue to themselves and to the country, than from the great merit and encouragement due to Mr. Bell.
>
> I had the pleasure of obtaining some subscribers for him, among others Neil Macolm, Esq., of Poltalloch, paid a £50 share, and I rather think his lady did the same. . [37]

Bell had, from 1814, an interest in another steamer variously called the *Stirling* or the *Stirling Castle*. This vessel was built in 1814 by John Gray at Kincardine on Forth for a consortium of Stirlingshire businessmen.[38] The *Stirling* had a somewhat chequered career, being fired on by musket or fowling-piece in August 1814, (though whether this assault was by a short-sighted wild-fowler or a rival ship-owner is not made clear.) More worryingly the *Stirling* was the subject of a court order prohibiting it from plying on the Forth in 1819. Her cast iron boiler had exploded and press reports argued that the cause was her adherence to a one-man operated engine room as opposed to the more normal complement of an engineer and a fireman.[39] In this context it is interesting to note that *Comet*'s engine room complement on her service between Glasgow and Fort William was a fireman and two enginemen.[40]

After working on the Firth of Forth between Stirling, Alloa, Newhaven and Dysart for about six years the *Stirling* changed hands, although retaining Bell's involvement, and transferred north to commence a service on the Caledonian Canal. Even before this move was made she had been integrated into Bell's West Highland services. In April 1820 Bell had placed large advertisements for the *Stirling* and *Comet* steamboats in the Glasgow press. These gave *Stirling's* timetables for her Firth of Forth services and pointed out how convenient they were for travellers from Glasgow to Edinburgh who could use the Forth and Clyde Canal passage boats to and from Lock 16 and take the *Stirling* on into Edinburgh. The advertisement went on to note:

> The COMET Steam Boat is appointed to sail from Glasgow every Thursday morning for Fort-William . . . so that by these two conveyances the public can be carried from the East Coast to the West Highlands, and from thence back to the Capital of Scotland in a most cheap and expeditious manner. . . By this mode of conveyance the public can be carried from Edinburgh to Fort-William for Thirty-four shil–lings, including coach hires from Edinburgh to Newhaven, and from Grangemouth to Lock No. 16.—Steerage passen–gers, Twenty shillings.[41]

Bell, with his flair for advertising and obviously being an assiduous promoter of the delights of travel to the Highlands, did not fail to add that:

> The beauty of natural scenery on the passage, is not to be equalled in Europe.

Joseph Mitchell[42], who as a young man working in the Caledonian Canal office describes sailing on what one takes to be *Comet II* on her maiden trip eastwards through the Canal from Fort William, says that what he calls the *Stirling Castle* worked initially East to West, from Inverness down as far as Fort Augustus. After 1822, when the Canal was opened from sea to sea, a through service was run to connect with *Comet II* plying from Glasgow to the Atlantic end of the Canal at Fort William, although the honour of the first complete passage of the Canal— from sea to sea—went to *Comet II*. Mitchell unfortunately cannot always be entirely relied upon as a witness, he being one of the many writers who manages to confuse the two *Comet*s.

From other evidence it seems clear that *Stirling* had been put on to this Canal service even during the lifetime of the original *Comet*. Sir Hugh Innes had written to Seaforth in May 1820:

> ... Mr. Bell is now at Fort Augustus or Inverness making arrangements for a boat from Inverness to Fort Augustus and a carriage from there to Fort William.[43]

and it seems apparent that Henry Bell's fertile mind had early on seen the great advantages of being able to offer the through journey from Inverness to Glasgow and indeed, from there, though no longer on his ship, onwards to Edinburgh.

Although the *Stirling* was not to change her Port of Registry to Inverness until 1824 she was certainly in service on Loch Ness at an earlier date. When she was registered at Inverness her new owners were a mixture of local interests, such as James Murray Grant of Glenmorriston and Andrew May of Clachnaharry, and professional contacts of Bell's, including Hugh & Robert Baird, Duncan MacArthur, and perhaps most interestingly of all, the great Thomas Telford, the consultant engineer and designer of the Caledonian Canal. Telford owned 2/64th of the *Stirling* while Bell's holding was 4/64th.[44] Edward Morris in his biography of Bell describes Telford as an old friend of Bell's and it is interesting to have the joint interest in the *Stirling* steamship as at least a degree of confirmation of some sort of relationship between Bell and Telford, one of the great civil engineers of that age of great engineers.

The situation which is revealed by the registration

documents, namely, a wide share ownership among local people and general investors, was evidently not easily arrived at. Although in July 1820 Alexander Anderson wrote to Seaforth from Inverness about Bell's activities and could report:

> Bell has got a great proportion of his Fort William steamboat sold in shares of 50 guineas and I understand the concern is paying well.

(The reference to the "Fort William steamboat" is of course to the *Comet* and confirms the evidence of Bell's correspondence with Dr Wright and his comments to Seaforth on the profitability of the service.)

Anderson goes on to discuss the sale of shares in what must be the *Stirling*:

> He values the one now plying on Loch Ness at 2000 guineas and candidly owns that unless he can dispose of from a half to two thirds of her in shares of the same description so as to lighten his advance and interest the country at large in her prosperity he will be obliged to look out for another scene of action.

From this letter we may deduce that Bell has borrowed a substantial sum of money—"his advance"—to buy out the rest of the original Stirlingshire owners and to place the *Stirling* on the Loch Ness service. Anderson concludes:

> Although I am neither a commercial man nor a landholder I should be exceedingly sorry that this should happen, as the interest of the northern district would certainly be much promoted by opening up this internal communication and if the canal was once opened to Fort William I am satisfied that there would be no loss by the concern—on the contrary a very fair prospect of its turning out a profitable one. So far as my limited influence extends I am doing all I can to procure him shareholders, but an unfortunate apathy appears to

prevail among those who should be most forward in the undertaking.[45]

It is worth noting that Anderson's judgement "that there would be no loss by the concern" was to be proved sound. In November 1823 the *Glasgow Courier* would report that:

> The Loch Ness Steam Boat has this season been so successful as to pay off all debts and divide 12½ percent.[46]

The apathy that Anderson describes may well go a long way to explain Bell's comments in his August 1820 letter to Seaforth about the reluctance of the Highland landed interest to invest in such projects.

Joseph Mitchell gives another example of Bell's activities in this field when he tells how:

> In the summer of 1825 Mr Henry Bell of Glasgow called on me in regard to steam communication he was then about to establish between Glasgow and Inverness with a coach to Skye. It seems that my father had been a shareholder in his *Stirling Castle* steamer, then plying to Fort Augustus. Bell dined with me, and we became confidential friends.[47]

If Mitchell's memory can be relied on as to the date of his meeting with Bell then the passage presents a slight problem. Mitchell elsewhere states that Bell had opened up the Glasgow to Inverness route three years earlier in 1822. This date is confirmed by Bell's memorial to Robert Bain, the captain both of *Comet* and *Comet II*, which states quite categorically that Bain was " the first Captain who commanded a Vessel from Sea to Sea, through the great Caledonian Canal, in 1822" and also by a list of canal traffic printed in the 20th Annual Report of the Caledonian Canal Commissioners which shows for 28th November 1822 the *Comet* Steam Packet, Captain Bain, carrying passengers from "Western to Eastern Sea" and on 29th November making the return trip

"Eastern to Western Sea"—the first such entry in the list.[48] What we may safely take out of Mitchell's reminiscence is not that the Glasgow to Inverness connection was first opened up by Bell in 1825, but that a new element in his plans dates from that year, which would seem to have been the coach connection onwards to Skye. This would again have been an attempt to serve a highly attractive and increasingly popular tourist route and also a strategy to compete with rival vessels such as the *Highlander* and *Ben Nevis*. Joseph Mitchell's statement about his father's investment in the *Stirling* is confirmed by the Inverness registry documents which show John Mitchell as a 2/64th part-owner of the vessel.

The popularity of steamer sailings for tourists had long been clear to Bell and to others and had been conclusively demon–strated on the Clyde and neighbouring waters. Back in 1820 a Greenock firm had written to Seaforth about the possibility of a steamer service to Lewis and had commented:

> There exists here and Glasgow, we may call it a fanaticism for steam boat excursions in summer which we have no doubt would make the present plan pay remarkably well. The novelty of seeing the Hebrides and so easily would attract multitudes to all parts of the North of Scotland.[49]

In 1823, with the *Comet II* well established, an article in the *Glasgow Courier* echoed this enthusiastic report about the commercial potential of steamships. The writer suggested the development of Oban as an interchange point for steamer services between the Clyde and Inverness, with one steamer working from the Clyde to Oban and another from Inverness to Oban via the Caledonian Canal and Fort William.

Arguing that this, together with the use of Greenock as a point of departure rather than Glasgow, would allow two services per week from Glasgow to Inverness, the writer went on to argue for a third steamer based on Oban to serve the tourist trade to Skye, Iona and Staffa:

> . . . such a plan would soon turn out to be a source of profit . . .
> for it is well known that there are no steam-boats in the
> Kingdom better employed than those at present plying to
> the Highlands.[50]

Bell himself was not behind-hand with such touristic schemes.
On one of her April 1820 sailings *Comet* had fitted in a cruise up
Loch Eil from Fort William and Corpach.[51]

Despite his early and clear vision of the place that the
steamer would have in local and international trade, and the
obvious profitability of the West Highland services, Bell's luck
with his ships was sadly not of the best. The *Stirling* was wrecked
in Loch Linnhe a few miles south of Fort William, while *Comet II*
sank, with considerable loss of life, in a collision with the steamer
Ayr off Greenock on 21st October 1825.

Quite how much responsibility Bell had for the operation
of *Comet II* at that date is somewhat uncertain. The date he ceased
to act as Superintendent for the Comet Steam Boat Company is
not known. However his memorial to Bain speaks of Bain's "16
years faithful service" to Bell—a period which would have
extended from 1814 to Bain's death in 1829. Robert Bain however
was not in command of *Comet II* on the day of her loss and he
may have been employed elsewhere on another ship in which
Bell was interested. In any event Bell certainly seems to have
been sufficiently distanced from any involvement in or
responsibility for the operation of the steamer to feel able to write
a lengthy letter on the subject of the collision and of safety at sea
to the *Glasgow Mechanics Magazine*[52].

His proposals are for the most part well thought out,
comprehensive, sensible and cogently argued. They include
legislation to license steamboats; a restriction on the number of
passengers to be carried, this restriction to be linked to the engine
power of the vessel. He called for experienced seamen to be
appointed as captains, pilots and mates and for certificated
engineers. In this Bell was considerably ahead of his time—the
examination of the competence of masters and mates only

THE WRECK OF THE COMET,

The Comet II *in collision with the Steamer* Ayr, *off Greenock, October 1825.*
(Dumbarton District Libraries)

commenced on a voluntary basis in 1845 for the foreign-going trade and was not made compulsory for home-trade officers until 1854. Engineers were not to be brought within the scope of such legislation until 1862, nearly forty years after the loss of *Comet II* and Bell's proposals. His letter also argues for an inspection system for steamboats, their engines and machinery with a general inspector and deputies at the various ports around the country—a structure in fact somewhat similar to the system of inspection and certification by Lloyds of London which was eventually to come into force.

The somewhat haphazard and ill-regulated nature of sea-faring in 1825 is indicated by Bell's fourth proposal:

> That those steam vessels be at least furnished with two lights, one on the bow, and one at the mast-head to be put up at one hour after sun-set, and properly attended to; also an alarm-bell, at night, attached to the engine, and a proper watch kept a-head with a speaking trumpet to direct the steersmana of the vessel.

That such a proposal, which perhaps seems to us now to be no more than the most basic common sense, was then badly needed cannot be doubted. Bell notes that the *Comet / Ayr* collision:

> . . . arose entirely through carelessness

and points out the great need:

> . . . that some steps should be taken to prevent in future, as far as possible, such accidents occurring through carelessness or mismanagement.

Bell, being Bell, does not fail to point out his qualifications for rendering this advice:

> As I have had the honour of bringing steam vessels into

206

practice in Great Britain, as well as other countries, I beg
leave to suggest what would, I conceive, be an improvement,
and be for the safety of the lieges.

and he expresses the hope that:

> ... the few hints I have thrown out, or such like, will show
> the public the necessity of some better regulations being
> immediately adopted.

Whether due to Bell's "few hints" or not, regulations along these
lines were eventually to be introduced. Not all his proposals,
however, became accepted—after suggesting that steamboats
meeting each other give way to the port side, he went on to
propose that on being overtaken they:

> ... do the same, and allow them to pass by on their starboard
> side, by stopping their engine as soon as the one over-taking
> them comes within 30 feet of their stern; and all sailing-
> vessels to give a sufficient birth for steam boats passing with
> freedom. . .

which is perhaps a natural enough sentiment in the man who
fathered steam navigation in this country but was a view which
failed to win acceptance and the general rule for priority between
sail and steam at sea was to be enshrined in the phrase "steam
gives way to sail".

The wreck of the *Comet II*, which in William Thomson's
colourful phrase:

> ... a penny candle in the bow of each steamer might have
> saved.[53]

resulted in the loss of at least sixty lives. *Comet II* was completing
a journey from Inverness and between two and three o'clock in
the morning was just off Kempock Point near Gourock when

she crossed the path of the *Ayr* steamboat making passage from the port of Ayr to Greenock. The *Comet II* was not showing any lights, although the *Ayr* was. The *Comet* seems to have kept a poor look-out and the two ships were so close to each other that when her look-out spotted the *Ayr* his call to the *Comet*'s pilot to put his helm a-starboard, was, according to a *Greenock Advertiser* report[54], heard by the *Ayr*'s pilot and taken to be an instruction to him. The two vessels, which were sailing at their best speed, turned into each other and the *Comet II* sank rapidly.

The loss of life was made worse by Captain Thomas McClelland of the *Ayr* immediately leaving the scene of the accident, apparently in the belief that his own vessel was in danger of sinking, and abandoning the survivors struggling in the water.

The true extent of the *Ayr*'s damage is perhaps best indicated by the fact that she was able to sail from Greenock on Saturday 22nd October, the day after the accident. The surviving senior officers of both ships were arrested and the master and pilot of *Comet II* stood trial, with Captain Duncan McInnes being sentenced to three months imprisonment.

The exact loss of life resulting from the sinking of *Comet II* seems to have been hard to establish, because of the difficulty of determining the true numbers of those on board, an unscheduled call having been made at Rothesay and an unknown number of passengers landed there. Fifty bodies of those drowned were recovered in the following week but one newspaper report commented:

> It was formerly stated that 47 were accounted for as drowned and 13 saved; since that three more have been added to the melancholy list of sufferers, making the whole accounted for, independent of the missing, 63. The earliest announce-ment of this awful calamity seems to have been nearer the truth than any later accounts. It was then asserted that from 80 to 90 found a watery grave; and from those accounted for, and those still missing, we are sorry to say that this number has not been overstated.[55]

The press reports of the loss of the *Comet II* offer an interesting insight into the effects that the coming of steam navigation had had on the Highlands and on the ease and possibility of travel for all classes of society that the new steamer services provided. The list of the dead included John Bell, a flesher (or butcher) from Dumbarton, a private soldier of the 45th Regiment; Anthony Gallacher, an Irish pedlar; Duncan McKenzie, a Highland trader; Mrs McDonald, cook to Mr McDonald of Arisaig; Captain and Mrs Sutherland of the 33rd Regiment; Donald McBraine, shoemaker of Glasgow; an unnamed old man from Crinan; and, perhaps most poignantly:

> A stout young woman, unknown; had on a brown bom–basine gown, coarse grey worsted stockings, and shoes tied with white tape, supposed to have come in at Oban, and then to have had on a black bonnet and two black feathers.[56]

When, around the middle of the century, famine struck the Highlands, at the same time as the Great Hunger in Ireland, the destitution and suffering along the West Coast and the Islands, grave as it was, was eased by the existence of an efficient system of steamship links. This communication system not only enabled news of the situation to reach the South but made possible the transport of relief supplies to the Highlands and Islands.

The origins of this vindication of the steamship long after Bell's death and the growth of the Highland steamer network are to be seen in the pioneering voyages of the *Comet*, the *Comet II*, the *Stirling Castle* and their rivals on the West Coast trade such as the *Argyll*. The *Argyll's* owners made a point of emphasising their Highland connections by including in their newspaper advertising the Gaelic slogan "Uail na Cluaidh a's Deo-Ghreine nan Gael"—the Pride of the Clyde and the Gael's Ray of Sunshine.[57] These vessels were indeed the pride and talk of the river and by their role in opening up the commerce of the West Highlands could, indeed, be seen as shining a new light into what had been in many respects a closed and inward-looking society. Bell's confidence in the transforming power of the steam-engine and the part that the steamship would play in uniting the world proved to be as fully justified in his own home waters as in the more spectacular arenas of trans-Atlantic and long distance trade.

Henry Bell's shipping interests extended beyond the *Comet*, *Comet II* and the *Stirling*. In September 1822 the *Highland Chieftain*, a 77 foot fifty three and a half ton steamboat was re-registered on change of ownership.[58] Listed among her shareholders was "Henry Bell, Engineer in Helensburgh". Bell joined a varied cast of proprietors in ownership of this frequently re-registered vessel. *Highland Chieftain* had started her life trading exclusively on what was considered as the inland navigation route from Dumbarton to Glasgow. At this period she was known as the *Duke of Wellington*, under which name reference was made to her in Chapter Seven. She was then, after extension and repairs, registered for the first time as a sea-going vessel at Glasgow and re-named *Highland Chieftain*, before being transferred to Dumfries and re-registered there in May 1821.[59]

A press reference to *Highland Chieftain* has already been quoted indicating that she had been in service to Skye and the Outer Isles in late 1820. However at that time Bell was not listed as one of her owners. The 1822 ownership list of the vessel is of some interest in that it indicates how widespread investment in steamships had now become, including as it does a vintner, a grocer, a baker, a merchant, a smith and a lawyer, all residents of Glasgow.

The *Stirling*'s career ended in January 1828 in a "perfect hurricane" while on passage down Loch Linnhe *en route* from Fort William to Oban. Loch Linnhe was not the luckiest of areas for Bell—the site of the *Stirling*'s wrecking is only some four miles from Sallachan where *Comet I* had run aground on her final voyage. A gentleman who was resident on the Ardgour shore of Loch Linnhe near the scene of the wreck sent a swift report of the accident to the ship's agent in Glasgow. His letter was published in the *Glasgow Courier* and it will be noted that the confusion over the title of the vessel continues, with the newspaper headline using the name *Stirling Castle* and the text of the letter the form *Stirling*.[60]

Melancholy Accident
Loss of Stirling Castle Steam Boat
Innerscadell by Appin
7th January 1828

Sir,

I am sorry to be obliged to intimate the melancholy account of the loss of the Stirling Steam-boat, under my house, six miles after leaving Fort William this morning. As far as I can learn, every exertion on the part of the captain and crew had been done to save the vessel and passengers, but the gale had been such as rendered every exertion vain. All on board have been got on shore in life, except one man, an Englishman, who had been a butler with Mr Macdonald of Clanranald, and on his way home to Edinburgh, as Mr Macdonald of Glengarry, with his two daughters were on board, on their way to Edinburgh. Mr Macdonald had been so much hurt on landing, that he died at nine o'clock this night, in consequence of a contusion on the head. All the rest of the crew and passengers are in a fair way of recovery; all that was possible has been done for the comfort of the shipwrecked sufferers, as well as the saving of property. The vessel is a total wreck.

H. McDougall

Mr. John Laird, Agent, Glasgow.

Later press reports amplify and to some extent correct Mr McDougall's letter. One of the main landowners in Lochaber, Macdonnell of Glengarry, had boarded *Stirling* with his two daughters at his home, Invergarry House, by Loch Oich, on Wednesday 16th January. The ship had been delayed overnight at Banavie Locks on the Caledonian Canal and had sailed for Oban the next morning in very adverse weather conditions. Six miles out of Fort William part of the machinery failed and without engine power the ship drifted onto the rocks. The passengers

211

were ferried ashore but Glengarry, becoming concerned about the safety of his second daughter leapt from the ship into the sea, struck his head on the rocks but was able to swim ashore and, with his two daughters, make his way to Mr McDougall's house at Innerscadell. Once at Innerscadell, or Inverscaddle as it now appears on the Ordnance Survey, he at first made light of his injuries but eventually, as McDougall reported, died at nine pm. The only other fatality was the English butler. This casualty was not as McDougall indicates in the service of Clanranald but rather from the household of the Laird of Moidart. The site of the loss of the *Stirling* and Glengarry's fatal accident in Inverscaddle Bay is still distinguished on the larger scale Ordnance Survey maps by a half-tide rock bearing the name "Glengarry's Rock" or "Sgeir Mhic ic Alasdair" from the Gaelic patronymic of the Chiefs of Glengarry—the son of Alasdair.

There is a condsiderable irony in Macdonnell of Glengarry being killed by an accident on board a new-fangled steamship. Colonel Alexander Ranaldson Macdonnell, 15th Chief of Glengarry, a totally anachronistic figure, was perhaps the last of the old-style Highland Chiefs who maintained and kept up the traditional semi-feudal way of life on his extensive estates and kept up the traditional mode of Highland dress and manners. Alasdair Macdonnell, to use the Gaelic version of his forename, is generally accepted as having provided Sir Walter Scott with the model for the character of Fergus McIvor in *Waverley* and he was painted in full Highland dress and arms by Sir Henry Raeburn in a magnificent portrait "Colonel Alasdair Macdonnell of Glengarry" now in the National Galleries of Scotland.

There is yet a further irony in the circumstances of Glengarry's death. Despite the fact that the Chieftain had, as a leading local landowner in Lochaber, been one of the main participants at the official opening of the Caledonian Canal in October 1822 he nonetheless kept up a campaign against the Canal on the grounds that it invaded his privacy. In 1823 he wrote that "the privacy and retirement" of his mansion overlooking the Canal route down Loch Oich was damaged now that "passage boats and smoking steam-vessels" were regularly passing through Loch Oich and he demanded compensation for the loss of amenity this caused. [61]

One of the "smoking steam-vessels" was of course the *Stirling*, which, despite his wish for seclusion he clearly did not refuse to use. Indeed, for all his protests, Glengarry did not disdain to make a profit out of the Canal—he sold land at Loch Oich to the Commissioners for canal building purposes for £10,477.7.1 and between 1804 and 1816 sold £3347.8.3 of timber to the Canal.[62]

It is said that the purpose of Glengarry's last journey in January 1828 was to go to Edinburgh to try to put his somewhat chaotic affairs in order. Certainly his heir, at the time of the accident a schoolboy at Eton, afterwards found it necessary to sell the family estates and emigrate to Australia.

It is perhaps not overfanciful to see in the microcosm of Glengarry's life and death the process of change in the Highlands, a process which was undoubtedly accelerated by the advent of rapid and, despite the tales of shipwreck recounted above, reliable steam navigation. A century before Bell's era the building of a network of military roads by General Wade and his successor Major Caulfeild in the aftermath of the 1715 Jacobite rising had opened up the Highlands to centralising influences and signalled the beginning of the end of the power of the clan chiefs, a process memorably recorded in Neil Munro's novel *The New Road*. Similarly in the early nineteenth century the coming of Bell's steamers allowed the swift movement of people and goods around the Western seaboard to the benefit of the economy of the region.

One body of officials who observed this process were the Caledonian Canal Commissioners who, in their 19th Annual Report, issued in 1822 noted that:

> . . . a steam packet plies regularly between Inverness and Fort Augustus and had found or created much more intercourse than could have been anticipated.[63]

In the following year the Commissioners were able to celebrate the opening of the Canal from Sea to Sea and noted in their report that:

> . . . a regular passage between Inverness and Fort Augustus was established by means of steam boats in the year 1820 . . .[64]

and went on to describe the opening of the Canal and to look forward to future developments in steam navigation on the Canal:

> . . . steam boats will be placed in the Lakes as soon as there is demand for their services in towing heavy vessels.

A very full account of the opening of the Canal and the attendant celebrations appeared in the *Inverness Courier* of 31st October 1822 and was reprinted by the Commissioners as an Appendix to their Report. Among other comments it was noted that the official party were joined on the second day of their celebrations—Thursday 24th October—by:

> . . . the *Comet* steam-yacht, which had politely been sent by the proprietors to meet the Commissioners and tender its services. . .

At a dinner that night in the Masonic Lodge in Fort William sixty seven gentlemen, including Sheriff Substitute Robert Flyter of Fort William (in his spare time the Treasurer of the Comet Steam Boat Company), but not apparently including Henry Bell (not perhaps a gentleman), sat down to a "handsome and plentiful dinner" and worked through an interminable list of speeches and toasts which concentrated on the economic and social benefits to be derived from the newly completed canal.

Just as the landowners of Inverness and Ross-shire applauded the benefits of the canal so the development of steam navigation won the general approval both of improving landowners like Seaforth and Innes and mercantile interests both in Lowland Scotland and in Highland commercial centres such as Fort William and Oban. All of these groups were quick to see the commercial and social advantages of the steamer and to take up shares in the steamboat companies.

The impact of the steamship on the general population was also marked. The integration of the Highlands into the economy of Lowland Scotland was, to a large extent, accelerated and facilitated by the coming of the steamboat. Seasonal migration to find employment as harvest workers on lowland farms had long been a feature of the area but this became more important and much more easily arranged with the rapid and regular shipping services offered after 1819. One indication of the extent of this steamer traffic came in 1827 when the *Inverness Courier* reported that in two weeks of August:

> ... upwards of 2,500 Highland shearers passed through the Crinan Canal for the South, in the steamboats *Ben Nevis, Comet* and *Highlander*, from the islands of Mull, Skye, etc.[65]

Equally, by opening up the Highlands and Islands to visitors, to outside ideas and to news, and by making the intervention and operation of central government easier and more effective, the new steamboats also tended to hasten and confirm the decline of the traditional systems and power structures in the Highlands which had survived largely as a result of the area's geographical isolation.

The era of the traditional Highland chieftain, like Macdonnell of Glengarry, living on and from his estates, surrounded by an elaborate hierarchy of feudal retainers and measuring his wealth and power by his tail of warlike followers was over. The failure of the Jacobite Rising of 1745 had confirmed the inevitability of this. In the nineteenth century the Highlands were to become increasingly a place where glens, which had once carried a large population, were cleared and became filled only with flocks of sheep and a few, often Lowland, shepherds.

Sporting visitors from the South increasingly replaced the indigenous aristocracy, a process symbolised by the discovery of the Highlands by Queen Victoria and Prince Albert and given form by their purchase of the Balmoral estate. Both changes were reliant on a revolution in transport to make them practicable. The depopulation of the Highlands and the economic transformation which was to result in the feeding of the growing population of industrial Scotland with Highland mutton and herring were alike facilitated by the transport revolution ushered in, back in 1819, by *Comet* passing through the Dorus Mor and penetrating the waters beyond Crinan.

CHAPTER 9
"His 'Ruling Passion' Scheming"

It is unlikely that it will ever be possible now for us to get a full picture of all Henry Bell's multifarious activities, schemes and ideas. Margaret Bell apparently preserved a large archive of her husband's papers relating to his many projects, some of which were available to and used by Morris in the preparation of his biography. Sadly these papers do not appear to have survived, doubtless due to the lack of close family or a direct heir when Margaret died in 1856. However the surviving scraps of information that are available to us, either through the medium of Morris's work or preserved by chance in libraries and archives, remind us of Bell's range and give a tantalising impression of a wide-ranging imagination and a fertile brain. An undated fragment in Bell's hand, preserved in the National Library of Scotland's Paton Collection of Manuscripts, discusses the possible deepening of the River Leven between Loch Lomond and Dumbarton and is so typical of its author's style that it is worth quoting in full:

> ... of the improvement of navigation of the watter of Leven could be so seafly obtained at so littel expens and that the propryters would have at least 5 to $6\frac{1}{2}$ percent for their outlay beside each public work on the Leven improved by gett at least 2 feet of more fall in their works beside the whole use of the watter wheir in grate drouth they have not one third of it.[1]

This fragment is endorsed "The above is from the pen of the late Henry Bell—a slip from the back of a drawing of his" and is signed by Edward Morris.

What Bell is proposing in this fragment is evidently a deepening of the River Leven—a scheme which would have had benefits for the many textile works situated along its banks. These bleach fields, print works and associated enterprises drew water from the Leven through numerous mill lades both to drive water mills and for processing purposes. Bell's interest in this industry and its requirements was far from abstract or theoretical. He knew the textile works of the Vale of Leven and their needs well. Not only was his Helensburgh home just a few miles away from the Vale of Leven but he had been employed to re-build the Dalmonach works there in 1812. Civil engineering works involved with water were a longstanding interest of his, and indeed were just what his apprenticeship as a millwright had qualified him to undertake. His proposals regarding the Glasgow water supply scheme were discussed in Chapter Three and we have also looked at the plans, during his Provostship, to provide his adopted town of Helensburgh with a good water supply. The Leven proposals thus would seem to arise from a combination of technical expertise and local knowledge.

The Paton fragment would seem to have been a draft for a proposal. To whom the proposal was to be addressed or whether it was ever sent is not known.

An even more ambitious proposal is referred to by William Maugham, the Dumbartonshire historian, who writes:

> He had also a scheme for the partial drainage of Loch Lomond, and reclamation of the land, about which he had an extensive correspondence with the Duke of Montrose, who did not receive it favourably.[2]

One may perhaps sympathise with the Duke.

Quite what the effect of Bell's proposal would have been on Scotland's most famous loch is hard to imagine. Maugham's story, however improbable it may at first sight appear, is not intrinsically impossible. The fact that it was to the Duke of Montrose that Bell addressed this startling proposal places the intended area of operation of his plan in the Montrose estate on the south east side of the Loch. The broad lower part of Loch Lomond is reasonably shallow, particularly in the area around the mouth of the River Endrick, and reclamation of this area would seem to have been the most likely target for Bell's scheme.

Loch Lomond, although fed by a large number of rivers has only one outlet, the River Leven, and one presumes that any scheme for drainage and reclamation must have, in part, involved increasing the outflow from the Loch through the River Leven. The possible convergence between Bell's Loch Lomond drainage scheme, as proposed to the Duke of Montrose, and the remarks on the deepening of the Leven in the Paton fragment is an obvious and attractive one, but in the absence of any further evidence must remain in the area of speculation.

Maugham also mentions that Bell gave thought to a Suez Canal, this around thirty years before De Lesseps was given the concession to build such a waterway. Maugham comments:

> Of all his plans he was exceedingly sanguine; neither the indifference of others, the want of resources, partial failure, or any of the thousand embarassments that haunt projectors, daunted him. Whatever the failure or disappointment met, he was always hopeful of ultimate success.[3]

By the 1820s there is ample evidence that Bell's health was deteriorating. In December 1819 in a letter to Sir Hugh Innes he apologises for not having written sooner but explained that he had been ill, his problem being:

> . . . in pairt watter and a groath which I had to be cutt for. . .[4]

Added to these specific problems seems to have been a chronic affliction in one leg. Donald MacLeod notes that:

While in his prime, Bell, engaged at some heavy work, had overtaxed himself, fell ill, and the illness resulted in a weakness in one of his legs. This leg for many years troubled him, and gave him a slight halt in his gait. . [5].

MacLeod goes on to tell what may be felt to be rather a tall tale about Bell's weakness in the leg.

The doctor assured him that so long as certain symptoms he indicated remained he would enjoy fair health, but whenever these ceased the result would be fatal. Very shortly before his death he called in an old friend to speak with him. He was then confined to his room, and said, 'R-, the doctor's prediction is about to come true, for my leg no longer troubles me.

Writing in June 1829 Bell, confined to bed, told Edward Morris:

The wounds in my legs are rather easier during the last few days . .[6].

and Morris in his biography notes:

For many years Mr Bell was lame and infirm, from injuries he had received in his arduous and enthusiastic efforts to perfect that mighty system which will immortalise his name . .[7].

which does coincide with MacLeod's assertion that the leg problem was caused by some form of work-related injury. Unfortunately we do not know the exact nature of Bell's problem and, dying before the era of civil registration, there was no death certificate recorded to show the cause of death.

Despite failing health the passion for scheming and his fertile imagination remained undimmed. In January 1820 he published a pamphlet entitled *Letter to the Honourable the Lord Provost of*

Glasgow; and Trustees on the River Clyde, containing the outline of a Plan for the Improvement of the Harbour, construction of Wet and Graving Docks, and Extension of the Navigation to Clyde Iron Works and Blantyre Mills.[8] It is worth noting that this elaborate, fifteen page proposal was produced at the same time as his being operated on for a growth, selling shares in the *Comet*, setting up the Comet Steam Boat Company, corresponding with the Highland landowners about the Hebridean steamboat and possibly carrying forward who knows what other schemes. Bell may perhaps be criticised with some justice for lack of method or for poor business sense but certainly not for lack of energy or effort.

The Clyde Trustees had applied for an Act of Parliament to authorise the construction of wet docks on the Clyde at the Broomielaw. Bell had seen a report of this and clearly felt that a more ambitious scheme should be presented to the Trustees to achieve their aims. He suggested that a 100 foot wide canal be built along the North bank of the Clyde from Stobcross to the Broomielaw and that the North wall of the canal be built sufficiently robustly to allow the construction of three or four storey warehouses on this embankment. From the main canal he proposed the building of branch canals at intervals of approximately 200 yards:

> The facilities for public works which these branches would afford, require no illustration . . . [and] . . . will besides, be found well calculated to serve for access to wet, dry, or graving docks. . .

This canal system would be at a level six feet above the level of the Clyde and entry to it would be by means of a 120 foot long lock, 30 foot wide, a length Bell considered sufficient for all anticipated traffic and, of course, wide enough for the wheels and projecting gangways of steam vessels. Such a canal would require to be fed with water and Bell offers the Trustees the choice of two schemes, the first and cheapest being a supply taken from the dam serving the lowest mills on the River Kelvin at Partick. This plan would, as it inevitably deprived the mills of their water supply from the Kelvin, require the purchase of these mills. However his favoured scheme involved nothing less than the erection of a dam across the Clyde at Glasgow Green, raising the water level there by three feet and conveying the supply to the Broomielaw by means of a:

> minor cut parallel to the north bank of the river.

Almost as a by-product of this secondary canal Bell holds out the possibility of using it to provide for barge traffic to the Clyde Iron Works at Tollcross or even as far up-river as Blantyre Mills— a distance of seventeen miles as the Clyde flowed in its winding path. Bell was anxious to emphasise that this proposal was no merely theoretical scheme and pointed out that he did not now propose to:

> . . . enter into the further details which have arisen from the minute survey which I have now completed.

Indeed his proposals are remarkably well detailed:

> Where the canal is to pass along the present Broomielaw quay, there will be found sufficient space to leave a street of 30 feet clear, between the north bank, and the present range of buildings. The canal will be secured by an iron railing, affording the most complete protection, not only to passengers, but also to all goods and property on board of vessels. . .

The aesthetic aspects of his proposals were not overlooked:

> Such a railing, independent of its utility, will form an object of ornamental beauty, and convert the street into an elevated terrace, bordered by a fine navigable canal, and somewhat

similar in appearance to the cities of Holland, or the beautiful
parade which overlooks the Avon, in the city of Bath.

Bell attempted to allay all possible fears and objections to his
plans. Above the site of his proposed dam he claims, perhaps
rather over-optimistically, that the river banks were so elevated
as to remove all possibility of flooding, even with an extra three
feet of water in the Clyde. The one exception, the low lying area
known as Fleshers' Haugh, would, like all meadows, he asserted,
be improved by being ocasionally laid under water.

He added that, quite apart from the fundamental harbour
advantages, the by-product of a greatly improved navigation
eastwards into the important Lanarkshire coalfields would be
of immense benefit to the city. He went on to advocate that a
horse-drawn railway be built to connect the Monklands canal,
which served a major coalfield, to his proposed works.

The costs of his scheme, including his preferred option of
the dam on the Clyde, would be £35,000 plus an additional £5000
for the extension of the navigation to the Clyde Iron Works. Bell
offers the Trustees his:

> ... best services in maturing and arranging the whole into
> such a form as will enable you to see the probable expense,
> and prepare you to carry the plan into immediate effect.

The old, ever-confident, Bell also can be seen in the closing
paragraph:

> It has ever been my study to promote objects of public utility,
> and I cannot help flattering myself with the hope of obtaining
> some share of your favourable opinion, when I reflect, that
> the system of steam navigation which I was the first to
> introduce, has added upwards of £2000 to the annual revenue
> of the river, without creating a single shilling of additional
> disbursement.

The Trustees did not adopt Bell's proposals.

223

Later however they did acknowledge the maritime and commercial community of the Clyde's debt to Bell by making him an annual grant of £50 for his support. This sum was raised in 1828 to £100 and this was continued to Bell's widow after his death in 1830.

In 1826 Edward Morris entered Bell's life. Morris, author of *British River of Death* and, as he describes himself on the title page of his *Life of Henry Bell*, "honorary lecturer of the Glasgow Temperance Societies", became acquainted with and interested in Bell's story, appropriately enough, through sailing between Glasgow and Liverpool on the *Henry Bell* steamer. Morris would seem to have taken up Bell's cause with energy and unbridled enthusiasm. Indeed his *Life of Henry Bell* is as much an account of Morris's energetic campaigns on Bell's behalf as it is a biography of the steamship pioneer. In 1826-27 Morris wrote numerous letters to the press about Bell's merits and organised collections and petitions on his behalf. While Morris was to become a close friend of Bell's it is significant that Morris's first letter, in September 1826, was written before he had in fact met Bell and was thus motivated only by an abstract consideration of Bell's merits and entitlement to public recognition, and in it he called on the community to support Bell's claims to a reward from the government. Bell, he argued, richly deserved the recognition of his government and had, with proper feeling, turned down ofers of support from other quarters, even from the commercial interests most closely associated with his work.

> A captain of one of the Clyde steamers told me some time ago, that the proprietors of the steam-boats offered Mr Bell the proceeds of a day's sailing of all the boats on the river, and this annually, which he refused to accept. I think he did right. It was kind in them to offer this; but Mr Bell looks to his government for reward. A boon from head quarters is what he richly merits, and will yet wait with hope that the British government will take up his case; with the determi–nation to return him something for all that he has done for

us, and not wait till his death, and then aid in the solemn
farce of erecting a monument to the memory of one whom
they neglected when he was among them. . .[9]

Bell's period was of course one in which rewards, grants and
pensions from the government were not infrequently given and
were even more frequently looked for. In the previous century
Dr Johnson in his Dictionary had, in characteristically contentious
style, defined a pension as "pay given to a state hireling for
treason to his country" but did not himself, some years later,
refuse to accept a pension of £300 per annum from George III.
The idea that someone, like Bell, who had benefitted his country,
should be given some form of recompense was thus well
established and, as can be seen from the letters of Patrick Miller,
even a landowning gentleman did not deprecate the idea of an
official reward. Miller's objection was to the reward's possible
destination.

Local support for Bell's case was not lacking. In January
1827 the Merchants House of Glasgow had petitioned the House
of Commons in the following terms:

> That your Petitioners, as representing the Commercial
> Community of the City of Glasgow, feel it their duty to
> recommend in the strongest manner to the favourable
> consideration of your Honourable House the Petition of Mr
> Henry Bell, of Helensburgh, in the County of Dumbarton,
> Engineer, who first succesfully introduced the practical
> application of the Steam Engine to the Navigation of the River
> and Firth of Clyde, from whence it has been since brought
> into universal use.
>
> That Mr Bell spent many years, and incurred great
> Expense, in gradually maturing his improvements. He is old,
> lame, and in narrow circumstances, and appears to your
> Petitioners to have an unanswerable claim to a liberal
> remuneration from his Country for the eminent advantages
> of his labours to all Classes of the Community.
>
> And therefore pray this Honourable House to take into

consideration Mr Bell's public services, and to allow him such remuneration for his great sacrifices to the general benefit as to your Honourable House shall seem suitable and proper.[10]

The Merchants House petition is of some interest in that it asserts that Bell "spent many years . . . in gradually maturing his improvements"; a confirmation of Bell's own account and that of John Robertson and a refutation of John Thomson's hostile account.

In December 1828 Bell wrote to Morris:

Dear Friend,
    Having been much disappointed, by the British government neglecting my claims so long, I consent to a subscription, as proposed by my friends in Glasgow . . .[11]

Morris states that Bell had in the previous year, during the brief Prime Ministership of George Canning (April to August 1827), travelled to London and had an interview with Canning on the subject of a grant from the government but to no immediate avail. However the Premier later changed his mind and ordered the payment of £200 to Bell from the Exchequer. Bell and his supporters were cheered by this evidence of official support but:

Mr Bell's leading friends in this city thought it best to delay lifting the money, and to make another effort to induce Mr Canning to use his powerful influence as Premier, to make this an annual sum for Bell's life.[12]

Unfortunately for the success of this plan, Canning died on 8th August and Bell's prospects of long-term government support faded. Indeed, in all the confusion attendant on the change of administration, the chances of Bell's receiving even the one-off payment of £200 appeared slim. Morris describes how a friend of Bell's (in fact it would seem that the "friend" was probably Morris himself) called at the Treasury a couple of years after

Canning's death to attempt to collect the £200 and how a Mr Cotton:

> ... went into an apartment, and searching among a large pile of papers, found a small slip of paper, in Mr Canning's own handwriting and holding it up said 'Sir, it is fortunate for your friend Mr Bell, that I have found this bit of paper, for if I had not found it, we could not have paid the money.' The next morning a £200 Bank of England note was received from the treasury, which in an hour afterwards ... was on its way to Helensburgh; and this was all that Mr Bell ever received from his government, for services of more value than many Waterloos.[13]

This payment would not seem to have been made until the summer of 1829 because Morris prints a letter from Bell dated 19th June 1829:

> My Dear Friend
>   I write these few lines lyeing in my bed, unable to sit up. But the letter you sent me, with remittance of £200 from the Treasury, a gift ordered by the late Mr Canning, will relieve my mind a little, and enable me to get Mrs Bell's house finished, and to pay the tradesmen. I was afraid I should not have got this £200, little as it is. The wounds in my legs are rather easier during the last few days, owing to my keeping close to my bed. I will write to you in a day or two more fully. I am your old friend,
>   Henry Bell.[14]

The reference to "Mrs Bell's house" is presumably to the property Margaret Bell and her mother had bought back in 1816.[15] The importance that Bell evidently placed on the work to be done to this house arose, it seems reasonable to assume, from a concern that Margaret should have a secure home in the event of her losing the tenancy of the Baths Inn. It will be recollected that the Bells had not been the owners of the Inn since 1810. Although Margaret was in fact to retain the tenancy of the Inn through various changes of ownership for many years after Henry's death, Bell obviously could not count on this. At the age of 62 and in deteriorating health he was evidently becoming anxious about his wife's future security.

If the government was slow to act and grudging in its response it is noticeable that those closer to the shipping and engineering world recognised Bell's merits, and conscious of his needs, acted to aid him. For example in January 1828 the Fifth Annual General Meeting of the Shipowners Society of Hull voted a £10 gratuity to Bell[16] and Edward Morris in his campaigning and fund-raising efforts reported the willing support he received from Isambard Kingdom Brunel, Thomas Telford and many other leading engineers of the day.

The Clyde Navigation Trust in 1829 raised its annual grant to Bell from £50 to £100 and Morris's fund-raising efforts in 1828–29 produced a total of about £500. The sad state of Bell's finances is indicated by Morris's statement that this subscription had unforeseen and unfortunate side-effects.

> Individuals to whom he had formerly owed money, which it was impossible he could discharge, harshly came forward with old-standing accounts, thinking to procure at once, on the subscription which Glasgow and other places contributed to the joint benefit of himself and Mrs Bell. In this unworthy object they were foiled by the firm and wise conduct of his true friends, Mr James Ewing, Provost Garden, Dr Cleland, Mr McGavin, and other benevolent and influential gentlemen, who stood by him in all his difficulties. . [17]

The idea of a host of creditors descending on the old and sick Bell on the news that he was, for once, in funds is indeed deeply unattractive. However a sympathetic thought may be spared for Bell's creditors: a group who were, as we have already seen, both numerous and frequently long-suffering. It is some indication of the tangled state of Bell's affairs that £200 from the government and £500 from public subscription did not clear his debts. These sums would require to be multiplied perhaps 20 or 25 times to equate to late twentieth century values.

Old age, neglect, poverty, harrassment by creditors, shipwreck, repeated disappointment all combined with an ever-increasing infirmity would be more than sufficient to drive many men to despair or to apathy. Not so Henry Bell. To the end of his life his mind, spirit, imagination and enterprise were as strong as ever. The obituarist who wrote that:

His 'ruling passion', scheming, was strong to the last . .[18].

was indeed stating no more than the truth. Bell was not a man to take things lying down and some of the rebuffs he suffered during the early part of his career as steamship owner clearly still rankled years later. In January 1827 he wrote, in his own individual style, to Walter Logan, the Superintendent of the Forth and Clyde Canal, about the current state of trade on the Canal Company's track boats and referring with evident bitterness to what he saw as the Company's unfair opposition to his operations with *Comet* and *Stirling* on the Firth of Forth some ten years earlier.

Thursday 11 Jany 1827

Mr Loagan

Sir I have had the pleasur of be a passanger in the Companys Track boats within this last tow months and in both times their was not a single cabin passanger—this you will say it is owing to the times. I denay the charge in pairt—you will reclect, the time I had my steam veshel on the Firth of Forth—your friend Mr C told the Company that if he had that fellow Bell off the ground the Ferry Company and Canals boats wold pay well after redusing the fare, from 3/- to 6 in cabin—after all that I shut them out and obliged that Company after Eighteen months opishoon [?opposition?]—the said loftie Company sent me a letter which I have as a tocken of trayofil [?triumph?]] that the said Company could wish I wold rais the frights back, to 3/- and 2/—so shon as I had gained my object, of trayonfil [?triumphing?]—I left the ground to themselves.—Now friend the said company is paid Back, in their own grounds—supose I was to putt a steam veshel at Grangemouth to run in two and a half hours for 2/- and 1/- and the Canal track boats at 2/- and 1/- each from No. 16 to Grangemouth 1/- and 6d this wold cut of a grate many passangers from each—but our friends know this from H. Bell that it is the only way they can gain back their credit if this my news were goen in to I am sertent it is the only way they will regain their credit and my advise is furder never as a Company to run Down an Individual or highten their fairs as they Did when the track boats was paying from 6 to 7 thousand per year at that time their greid spoiled the whole rost—but if the said Company take my advise for onse—I am sertent it will pay 2 to one they have at present -

    I remain
    a warm friend to
    all genrous herted
    friends for the poor-
    I am
    Sir your Most
    obt servent
    Henry Bell[19]

One would like to know what Logan made of this letter, which with its colourful imagery, "their greed spoiled the whole roast"; the anger and contempt which clearly possessed Bell when his plans for steam navigation were being frustrated, "the said lofty Company"; the uninhibited advice, "never as a Company to run down an individual"; the idiosyncratic spelling and the sense of the writer's ideas running ahead of his pen or his ability to structure his ideas; is entirely characteristic of Bell. Bell's troubles with the Forth and Clyde Canal Company were well in the past but the sense of indignation was still clearly fresh in his mind.

To the end, however, Bell could still look forward positively to new schemes as well as look back to old frustrations. On 23rd August 1830, less than four months before his death, Bell addressed an open letter *To the Gentlemen, Freeholders and Merchants of Argyleshire*. In this document he proposed the construction of a canal across the mile wide neck of land between East Loch Tarbert and West Loch Tarbert at the north end of the Kintyre peninsula. Bell was not the first person to note the advantages of this route—Magnus Barefoot, the son of King Olaf of Norway, had according to tradition dragged his Viking longships across the Tarbert isthmus in the 11th century and in so doing claimed the Kintyre peninsula as part of the Norse lordship of the Hebrides.

Bell pointed out to his readers that his proposed canal, the route for which he had surveyed two years before, would provide a safe and swift means of passage for vessels trading from the Clyde to the West Coast. He pointed out that his plan had the great advantage of not requiring any locks or drawbridges and could be cut in a straight line through solid rock. Such geological conditions might at first sight seem to be a disadvantage. In fact the Crinan Canal had suffered from the canal banks collapsing due to its line running through geologically unsound areas of sand and mud which proved to be unable to bear the weight of the canal banks.

The idea of a lockless canal, and the consequent speed of passage, would have been very attractive. The Crinan Canal with its fifteen locks could take a fairly long time to navigate. In the days before steam the horse haulage time was six to eight hours and steam power would only have made a modest reduction in this. Again, being lockless, the projected Tarbert Canal would not have had the restrictive size limitation that the Crinan locks placed on that canal. Bell's plans were for a ship canal 50 feet wide at bottom broadening to 60 feet on top and offering three possible depths with 15, 18 or 21 feet at high water. Such a canal would indeed have been a major sea-way—in Bell's enthusiastic words:

> It would certainly form one of the grandest openings in Europe; but its utility would be incomparably more important than its rural magnificence, and fascinating beauty, attractive as these would be.[20]

It would have been much wider than, for example the Caledonian Canal which could only take vessels of up to 36 feet in width. Bell's visionary proposal was close in size and capacity to the much later Manchester Ship Canal.

Bell noted that, after construction, there would be little maintenance cost and few outlays other than the toll collector's fees; the absence of locks and the rock-cut construction being responsible for this very desirable situation. His suggestion was that the canal would attract not only coastwise traffic but merchant vessels trading to America and the West Indies. He set forth a proposed table of charges covering everything from small rowing boats at 2 shillings and 6 pence to vessels over 100 tons, to be charged for at the rate of sixpence a ton.

The total expense, assuming construction to the deepest level, was £90,000, although the more modest plan for a canal offering 15 feet at high water, sufficient Bell noted for:

> ... the generality of vessels trading to and from the river Clyde . . .[21]

would only be £37,000. Bell suggests that a joint stock company should be floated to carry through this scheme, a suggestion which forms an interesting contrast to other contemporary canal projects in Scotland. The Crinan Canal had been started as a private venture but was only completed once public funds were made available; while the Caledonian Canal was from the beginning a public sector undertaking, one of whose motives had been to facilitate the movement of Royal Navy warships from coast to coast and a design constraint had been the need to lock through 32 gun frigates.

Bell claimed to be motivated only by concern for the welfare of his country:

> If these hints of mine for the benefit of the commerce of my country, and especially of the shire in whose prosperity you are more immediately concerned, should be approved by

you, and should issue in your taking steps necessary for the completion of the scheme, I shall consider my exertions in the business well rewarded.[22]

For once this type of disclaimer may be sincere. It is difficult to imagine that the elderly, sick and indeed dying Bell could have realistically planned to take any active part in the establishment of the proposed company or the practical execution of the plan.

The scheme is, like others of Bell's plans, a curious mixture of the shrewd and the naive. The plan for a canal across the Tarbert isthmus is entirely logical, the benefits of a lockless passage indisputable, the practicalities and costings doubtless well considered. Bell indeed states that he has checked his estimates with several persons experienced in the construction of canals. However the fatal flaw in his ingenious scheme was the Crinan Canal. This, less than fifteen miles away, was already completed and functioning. The idea that the traffic of the area, even with the exclusive custom of trans-Atlantic shipping which certainly could not pass the Crinan's locks, could support two canals—one at the north end of Knapdale and one at south end—or that it would prove possible to raise £37,000 let alone £90,000 for such a scheme is manifestly absurd.

When the Crinan Canal was planned the Tarbert line had been considered as an alternative. Undoubtedly the Tarbert route had its theoretical attractions as a short-cut. However the chosen line of cut, via Crinan, discharged west-bound traffic into more sheltered waters in the lee of Jura and Scarba. More importantly, use of the Tarbert route would have involved a passage down the narrow, eight mile long, West Loch Tarbert. This beautiful stretch of water has the great disadvantage of running South West—that is into the prevailing wind on the West Coast of Scotland. As a sailing ship cannot sail directly into the wind such a route would have made for a passage that would be frequently difficult and at times impossible. A heavily laden and unwieldy cargo vessel faced with tacking into the prevailing wind would have had all too little sea room in the confined waters of West Loch Tarbert to execute such maneouvres. Obviously the advent of steam reduced this problem but for many decades the bulk of traffic on the coast would still be carried by sail and the West Loch Tarbert passage was in itself a damning argument against Bell's scheme.

Even if we accept that with the coming of the age of steam there was the theoretical possibility of an answer to the West Loch Tarbert problem by providing steam tugs to tow sailing ships, then it is possible that Bell's scheme might have won support, had the Crinan Canal not existed. However, with a state-owned canal in operation—albeit one with substantial restrictions on size and with irritating locking delays in its passage—the notion of building a rival canal could be no more than a mere pipe dream. Edward Morris gives Bell's plan the following commendation:

> The above outline of an ingenious—probably, a rational and practicable scheme, proves to the reader the activity of that daring mind, which was soon to close its inventions in this world for ever.[23]

But perhaps even the ever-loyal Morris can be detected as letting just the slightest degree of doubt appear in his distinction between the certainty of his description of it as "ingenious" and the caution of his "probably a rational and practical scheme".

The attempted promotion of the Tarbert Canal was Bell's last ploy. His health was rapidly failing. As the *Glasgow Chronicle* obituary was to note:

> His constitution had been greatly broken down for many years, and his bodily sufferings were frequently very great.[24]

Right up to the end Bell was, true to his character, still embroiled in controversy and dispute. Donald MacLeod tells how some days before his death:

234

... there had been some law proceedings threatened against him in regard to a wall or shed he had removed, which a neighbour claimed as his private property, and the matter was in the hands of the Fiscal. 'I'm told,' he said, 'that the "beadles" are to be up from Dumbarton about the wall job, and will carry some o' ye to Dumbarton, but they'll no get me—I'll be past their power.[25]

His prediction proved all too accurate. On Sunday 14th November 1830, he died at home at the Baths Inn and was buried on Friday 19th November in the parish churchyard at Rhu (or as it was then spelled, Row), some two miles from Helensburgh.

Henry Bell's funeral saw, for once, full and ungrudging honour being given to the pioneer. The shops in Helensburgh and Rhu closed in tribute, shipping on the river flew their flags at half-mast; the *Waverley* steamer, owned and commanded by Captain Robert Douglas, a former employee of Bell's, fired minute guns as the cortege passed from the Baths Inn *en route* to Rhu.

The funeral was attended by 140 people including many of the local clergy, notables and gentry including Lord John Campbell (the future 7th Duke of Argyll) and only the inclement winter weather prevented, so Morris asserts, an even larger attendance.

In 1834, four years after Bell's death, the author of the Helensburgh section of *Fowler's Commercial Directory*, after noting that Bell was buried at Rhu observed bitterly:

... on my lately visiting the spot (August 1834) how was I surprised, nay, will it be believed?-that not a single mark of respect has been erected to the memory of the illustrious dead, to tell the stranger where his remains are deposited![26]

The writer went on to draw a pointed comparison between this lamentable neglect and the memorial Bell had himself provided in the same churchyard to the memory of Captain Robert Bain of the *Comet* and observed :

Not in this manner did he [Bell] treat worth and merit. . .

He goes on to assure his readers that the absence of a monument was not due to any lack of proper family feeling:

His widow, Mrs Bell, would ere this have erected a tribute of respect to his memory, had she not (which I can positively state) been given to understand that certain individuals wished to have the honour of doing so; but shame to our country, shame to our merchants and mechanics, for whom he has done so much; it will remain for ever an indelible mark of disgrace; the finger of scorn will be pointed at them, and they themselves held up to derision, if they allow any longer the remains of this illustrious individual to remain unhonoured.

In fact Bell's grave was to remain unmarked and unhonoured for a further sixteeen years, although a monument to him, overlooking the Clyde at Dunglass Castle near Bowling, was erected in 1839.

Bell died intestate and when in July 1831 his affairs were settled before the Commissary Depute of Dumbartonshire the inventory of his personal estate and effects was to total only £707-15-6.[27] Of this sum £610-16-3 represented household furniture, farms stocks, crops, spirits, wines, wearing apparel and all other moveable effects—a fairly modest total when one considers that it included the moveable furnishings, tenant's equipment and stock in trade of the Baths Inn. It was noted in the inventory that his widow had continued since his death in the management of the deceased's personal estate as executrix. Margaret with perhaps more business-sense than Henry, left her affairs in somewhat better order, making a will in May 1839 and revising it by a codicil dated June 1848.[28]

Henry and Margaret Bell do not appear ever to have had

any children. Certainly none survived to the time of Henry's death. Bell's brother Thomas, sometime the driver of his Glasgow to Helensburgh coach, from the evidence of Margaret's will had settled down as a farmer in Helensburgh, and he, his wife Ann and their son Henry, were the couple's closest relatives, in both the geographical and the family sense. Thomas had however died before Margaret made her will.

Ann Bell and her son, Henry, were left £75 and £25 respectively. Margaret's cousin, Isabel Hamilton, described as residing with her, was left a life interest in £200 with reversion to Isobel's sister Mary Hamilton. Another cousin, Margaret Watson, of Crossmyloof, Glasgow was left £40. Outwith the family her largest single gift was £200 to Rachael Walker "presently residing with me"—presumably the same Rachael Walker as is shown in the 1851 Census return as the Hotel's bar maid. The residuary legatees were five of her nephews and nieces; Margaret, Elizabeth, James, Christian and Janet Young.

Margaret's legacies reflected her long-standing and deeply held religious interests. These were remarkably ecumenical for the period, in that they covered a variety of denominations. In the original will these included gifts to the Dumbarton United Seccession Church, and to both the Independent Church and the Original Burgher Associate Congregation in Helensburgh. The bewildering variety of presbyterian dissent from the Established Church of Scotland was somewhat simplified by the formation, in 1847, of the United Presbyterian Church which embraced the Relief Synod, the United Seccession Church and the Burgher Congregations. Margaret's 1848 codicil reflected this and, withdrawing the earlier provision, provided for legacies to the Helensburgh United Presbyterian Church of nineteen guineas to the Church itself, the same sum to the Missionary Society associated with the church for propagation of the Gospel in Africa, a similar amount for propagation of the Gospel in Jamaica and ten pounds "for behoof of Weak Congregations" and five pounds for the poor under the charge of the Helensburgh United Presbyterian Kirk Session—a total of £74.17.0 in church bequests.

Margaret Bell had in fact been instrumental in securing the establishment of what became the Helensburgh congregation of the United Presbyterian Church. Various presbyterian dissenters in Helensburgh had worshipped in the Auld Light Burghers' church in Colquhoun Square. In 1839 the minister and congregation of this church had joined the established Church of Scotland, leaving no place of worship for those holding to the tenets of the Relief and Secession churches. To fill the gap Margaret Bell, herself a follower of the Secession Church, made a room available in the Baths Inn, and the attendance at this preaching station grew and eventually the worshippers became recognised as a congregation of the Secession Church in 1844 and, after 1847, of the United Presbyterian Church.

In her original will Margaret had appointed trustees and executors and revised this list in 1848. On her death the surviving trustees declined to act and an application had to be made, by parties with an interest in the estate, to the Court of Session for the appointment of a Judicial Factor on Margaret Bell's estate. This appointment was made in July 1856 and the factor, Robert Duncan Orr, a Helensburgh banker, succeeded by January 1858 in his task of winding up the estate. The sale of the household furniture etc situated within the Baths Inn realised £352.12.2¼; the horses, carriages, farm stock etc located there made a further £360.0.4½ while the contents of Margaret Bell's cottage, the "Mrs Bell's house" of Henry's letter to Edward Morris in June 1829, realised £139.15.1½ with a further £36 coming from rents due on this property. The whole moveable estate realised £961.10.1¼

CHAPTER 10
"Bell's Scheming
Got Britain Steaming"

The originality of Henry Bell's work on the *Comet* has been legitimately questioned. His character has equally been subject to criticism. John Thomson, his one time associate and later rival, the builder of the *Elizabeth* steamboat, writing in his *Account of a Series of Experiments* . . . published in 1813 as a counterblast to Bell's *Observations* . . . noted that:

> It is not merely in his Observations that Mr Bell has arrogated to himself the merit of this invention. In a late trip which he made to Leith with the *Comet* Steam Boat, he lost no opportunity of gratifying his personal vanity in this way. The public Newspapers also announced it in various shapes. This wonderful vessel was advertised to be exhibited in Leith Harbour, (something in the manner of an itinerant's show-box), for the inspection of the citizens of Edinburgh—and the accommodating, public-spirited, disinterested would-be inventor, was even willing to give the good folks of Edinburgh an apportunity of being carried out to the Roads, by the *power of Steam*, for a small pecuniary equivalent. Whether such a method was likely to impress the public with a high opinion of Mr Bell's talents, or the importance of his assumed invention, I leave to them, as well as Mr Bell himself, on mature consideration, to decide.[1]

Thomson's pamphlet was written to advance the view that he

239

was the true originator of the steam propulsion system used by Bell in the *Comet* and that Bell had taken unfair advantage of his work. Thomson clearly felt deeply ill-used and his corrosive bitterness shows through in his description of Bell, "the accommodating, public-spirited, disinterested would-be inventor", and in his putting Bell's exhibition of *Comet* on a par with with a showman's attraction. He goes on to say:

> In Mr Bell's return from his Quixotic expedition, he happened to fall in with a small coasting vessel near Dumbarton, which had been becalmed while coming down the river. Mr Bell, indeed, very kindly took the vessel in tow, and brought her into Greenock; but as if this deed was too meritorious to be allowed to make its way to the public through the ordinary channels, we find it pompously announced, the ensuing day in the Greenock Newspaper—and here, as in other cases, the name of Mr Bell, as the inventor, was not forgotten to be introduced.

The substance of this part of Thomson's complaint is that Bell was a shameless self-publicist who took every opportunity, fair and unfair, to promote his steamer and his invention and did not behave with the proper degree of gentlemanly restraint, modesty and circumspection. To a degree it is difficult to argue the point. Bell's braggadocio was a source of constant complaint from his rivals. Even the seemingly somewhat diffident James Taylor, who noted that he was not himself a pushing or lucky individual, described Bell's 1816 communication to the *Caledonian Mercury* as:

> . . . a most impudent and arrogant letter. . .[2]

Bell certainly lost no opportunity to assert his claim to be the pioneer of steam navigation. Bell was, perhaps, as much a businessman as he was an inventor or a pioneering engineer of the first rank. His *Comet* was essentially a work of consolidation and broke no new ground in theory or design. He had, as we

240

have seen, had the advantage of studying the *Charlotte Dundas* and the benefits of discussions with Symington, John Robertson and doubtless many others. Bell's accomplishment was that he applied what was already available knowledge, took the risk, attempted what better-trained minds had said was not feasible, and in the process showed better-funded and better-prepared imitators what could be done.

It is hardly surprising that Bell's success should have rankled with many. He had come from nowhere and achieved the goal that so many others had sought after. To compound his offence he did not have the grace and modesty to hide his light under a bushel but unashamedly attempted to capitalise on his work. Taylor and Patrick Miller Jnr were, of course, equally anxious to profit from the invention but as "gentlemen" rather than "players" they attempted to put a veneer of respectability on their claims. It is surely very significant that in Patrick Miller Jnr's lengthy, and somewhat unbalanced, correspondence on his father's role in the development of steam navigation he repeatedly categorises both Symington and Bell by their artisanal occupations and, thus by implication, by their lower social class. Of Symington he writes that he was employed:

... merely as a mechanic . . .[3]

while he describes Bell to Lord Melville, in somewhat dismissive tones, as having been:

... carpenter to Mr Fulton or his first steam worker . . .[4]

and to Archibald Constable when he wrote protesting about James Cleland's version of events, which had given more credit to Bell than Miller thought he deserved, he describes Bell as:

... the carpenter whom he so highly eulogises. . .[5]

Miller wrote more than once on what he described as "Symington

241

and Bell's pretensions to be considered the original inventors of steam navigation" . His account of the development of steam navigation appeared in the *Edinburgh Philosophical Journal* for July 1825. In an appendix to this article which appeared in the *Edinburgh New Philosophic Journal* of April/October 1827 he describes Symington as being considered by Miller Snr:

> . . . in the same light as he did every other labourer or tradesman he employed about his different vessels. . .[6]

Symington had responded vigorously to Miller's 1825 article, which was reprinted from the *Edinburgh Philosophical Journal* by the *Edinburgh Courant* newspaper. The accusation of being simply an operative was one which clearly hit home. In a lengthy letter to the *Courant* Symington quickly picks up, and attempts to refute, Miller's suggestion about his non-professional status and social class:

> I am described in the narrative as the *practical engineer,* from which I suspect it is intended to be insinuated that I was a mere *hammer and anvil man,* but it is well known that I never was in the practice of working with my own hand, except for the purpose of instructing the mechanics who laboured under me.[7]

Pursuing the same policy of applying a social stereotype, this time to Bell, Miller notes that he:

> . . . was originally bred a stone-mason . . .[8]

and goes on to quote verbatim a letter from Bell to John McNeill about Fulton. This letter is, it must be confessed, in Bell's best breathless style and appalling spelling:

> I was this morning favered with your letter and in ansur to your Inqueres anent the leat Mr Robert Fulton the Amerecan

ingenair his father was from Areshair but what plass or famlay I canut tell but his self was born in Amereca. . .

It is hard to avoid the conclusion that Miller's purpose in quoting Bell so literally, which hardly aided the reader's comprehension of Miller's article, was to drive home the message that Bell was no more than an unlettered artisan. Writing in an age which was highly socially stratified and in which class distinctions were of great importance, to label Bell or Symington in this way was no trivial matter. In effect Miller was saying that a man of this class and background could in no way be seen as the originator or creator of the steam navigation system.

Of course Miller's arguments in favour of his father's primacy in the field, however violently prejudiced he might have been against Bell, were not without an element of truth. The Dalswinton steamboat was indeed set sailing on Miller's Dumfriesshire loch twenty four years before the *Comet* entered the waters of the Clyde and the *Charlotte Dundas II* had been laid up for nine years after her trials before the *Comet* was so memorably advertised to sail "by the power of wind, air and steam". These experiments of Miller, Taylor and Symington had been vital pioneering steps and all credit must be given to them. However there had been no practical outcome from them and Miller and Taylor's work had stopped. Miller and Taylor, and Bell's other detractors could, with profit, have remembered the words of Sir Francis Drake:

> There must be a beginning of any great matter, but the continuing unto the end until it be thoroughly finished yields the true glory.[9]

Patrick Miller demonstrated the steamboat with brilliant success, and we have no reason to dispute Taylor's opinion that:

> a more complete, successful and beautiful experiment was never made . .[10].

Having instructed Taylor to communicate an account of the trials to the press, he then had the engine removed and kept as an ornament in his house. Their later experiments on the Forth and Clyde Canal were equally successful and were again the subject of a press release but again Miller failed to proceed with the development of the system.

Instead he devoted himself to agricultural improvements on his estate. Taylor claimed that he spent many years vainly waiting for Miller's interest in steam navigation to re-awaken. He goes on to say that around 1812, twenty two years after their first experiments on the Forth and Clyde Canal, and on hearing of the coming of the *Comet* he had:

> . . . waited upon Mr Miller and stated that now was the time, or never, to preserve the benefit of the invention. . .[11]

Taylor in the same letter revealingly remarks that what he and Miller had achieved he considered to be:

> . . . only as a part, and have lived in the faith and hope of some day coming forward with a more finished whole. But this my limited circumstances in life and want of friends have as yet precluded.

It is impossible to avoid concluding that in both Patrick Miller and James Taylor we have, for one reason or another, a remarkable inability or unwillingness to see the job through. Taylor's comment on the Dalswinton experiment is perhaps significant:

> After amusing ourselves a few days . . .[12]

It is indeed difficult not to feel that Patrick Miller was imbued with a fatally dilettante spirit. Equally one must conclude that Taylor was too much the subservient tutor to have the determination and courage to go off on his own account and put his experience and what were undoubtedly sound ideas into practice.

Whether or not either Miller or Taylor fully realised the true significance of their work at the time is a moot point. Miller's original motivation had of course been extremely limited—an auxiliary power source for ships in distress. His first involvement had been fitting handcranked wheels to multi-hulled vessels, an exercise which as Symington sharply pointed out had proved to be totally unsuccessful as had his espousal of multi-hulled ships.

> Mr Miller conceived the idea of forming double and triple vessels, that he considered better adapted than any other to be propelled by means of wheels, and he laid out a great deal of money in attempting to prove that his suggestion was of importance, but his experiments all failed, in so far that no double or triple vessel, such as he recommended to the world, has ever been used up to the present date.[13]

Symington observed that he had been on board Miller's experimental vessel during one of these less than brilliantly successful trials of manpowered propulsion.

> The truth is that Mr Miller's experiments were unsuccessful in the eyes of every one but himself, as many can declare who are still living, (myself among others,) and who laboured in the turning of the wheels. I can well sympathize with Mr Taylor, the tutor of his son, who exclaimed in the agony of his exertions, that he wished for a steam engine to assist them.
>
> The hint of the tutor, it would appear from Mr Miller's own narrative, is the first suggestion of the practicability of applying the steam engine towards the end he had in view; and the public will be better able to appreciate the effect of the admission when they are informed, that I was on board of the same boat, or at least I was on board on one occasion when the necessity of such assistance was generally admitted.

Taylor, it would seem from this and other evidence, probably saw somewhat further into the matter than did his employer. If he did, then so much the worse. For if his vision was clearer then it was fatally compromised by his reluctance to act decisively. The somewhat lame excuse about his "limited circumstances in life and want of friends" is hardly a very convincing excuse for what was, by the time he wrote to Huskisson in 1825, half a lifetime of inaction. William Symington, for example, was less well-educated and less well-connected than Taylor but managed to persuade the Canal Company and the Duke of Bridgewater to back his schemes and to achieve a measure of success with the full practical accomplishment of the dream of steam power at sea only eluding him by an unhappy coincidence of chances.

Henry Bell was perpetually financially embarassed and certainly no better placed in terms of social position, in Taylor's terms, but he had the determination, and perhaps the necessary lack of scruple, to make use of other men's ideas, to run into debt, to scheme and push and struggle to drive matters to a successful conclusion.

> . . . continuing unto the end until it be thoroughly finished yields the true glory.

Whatever the imperfections of Bell's work, however shaky his theoretical grasp, the unavoidable fact remains that with the *Comet* Bell had achieved the dream of practical steam power at sea and ushered in the new age. Despite the litigation commenced by Symington, the criticisms voiced privately and publicly by Patrick Miller Jnr and James Taylor, it is noteworthy that the majority of Bell's disinterested and informed contemporaries recognised his achievement. Those contemporaries who were in a position to make an informed judgement on these matters were united in recognising the worth of his contribution. Isambard Kingdom Brunel, whose comments on Bell were quoted in Chapter One, also described Bell as:

246

... the planner of the first practical steamer ...[14]

and contributed generously to Edward Morris's fund raising efforts on Bell's behalf, handing over his money with the comment that:

> ... the Government ... have missed a good opportunity by neglecting Bell but better the subscription than nothing. . .

Engineers of a slightly later generation also recognised Bell's worth. One such was Robert Napier (1791–1876), "the father of Clyde shipbuilding", who had started in business on his own account in Glasgow just three years after the launch of the *Comet*, built his first marine engine in 1823 and who had established himself as one of the leading marine engineers on the Clyde by the time of Bell's death. Napier was one of the signatories to the 1825 testimonial which confirmed that the propulsion system then used on Clyde steam-vessels was in all essential respects the same as that which Bell had introduced on the *Comet* a dozen years earlier. It seems only reasonable to conclude that an engineer of Napier's distinction and wide knowledge of the Clyde engineering world would be in an unbeatable position to have a shrewd idea of what exactly Bell had done and what his real merits were. A man like Napier is thus perhaps a more reliable witness to Bell's claims to recognition, reward and "true glory" than Patrick Miller Jnr. Robert Napier's judgement is perhaps to be relied on the more because he was the friend, cousin and brother-in-law of the long-suffering David Napier, the builder of the *Comet*'s boiler. He would thus know, from the inside, of the extent of Bell's financial irregularities and scheming but could equally see behind the self-promotion and assertiveness of Bell to make an informed judgement on the man's solid achievement.

Napier's views on Bell remained constant as did his support for schemes to honour him. Napier remedied the lack of a memorial to Bell by erecting a statue in Rhu Churchyard in 1853.

When the local newspaper reported the first steps in the erection of a monument to mark Bell's grave, alluding to Napier, it noted that:

> ... the entire expense of the statue is borne by a celebrated engineer on the Clyde, upon whom, almost more than upon any other living man, the mantle of the late author of steamboat propulsion seems to have fallen.[15]

and it may be felt that such a man would not have wished to be seen commemorating an imposter, a thief of other men's ideas or any sort of unprincipled scoundrel. By 1853 Napier's reputation was international and the recognition he gave to Bell's memory is as significant in its source as in its generosity. Nineteen years after erecting the memorial at Rhu kirkyard he again recognised Bell by becoming one of the major contributors to the monument to Bell on the seafront in Helensburgh.

Official reward may, for a long time, have eluded Bell, but a welcome official recognition of his contribution came in the *Fifth Report of The Select Committee appointed to inquire into the state of the Roads from London to Holyhead*... In this report to the House of Commons the Committee noted that they had commenced enquiring into:

> ... the important subject of Steam Boats.[16]

The Committee gave a brief account of the development of the steamboat in the course of which they noted:

> In 1801 Mr Symington tried a boat that was propelled by Steam on the Forth and Clyde navigation. Still no practical uses resulted from any of these attempts. It was not till the year 1807, when the Americans began to use Steam Boats on their rivers, that their safety and utility was first proved. But the whole merit of constructing these Boats is due to natives of Great Britain; Mr Henry Bell, of Glasgow, gave the first model of them to Mr Fulton, and went over to America, to

assist him in establishing them; and Mr Fulton got the engine he used in his first Steam Boat on the Hudson river from Messrs Boulton and Watt. . .

Mr Bell continued to turn his talents to the improving of Steam apparatus, and its application in various manufactures about Glasgow; and in 1811, constructed the *Comet* Steam Boat, of twenty-five tons, with an engine of four horse power, to navigate the Clyde between Glasgow and the Helensburgh Baths, established by him on an extensive scale. The success of this experiment led to the constructing of several Steam Boats, by other persons, of larger dimensions and with greater steaming power; these having superseded Mr Bell's small Boat on the Clyde, it was enlarged, and established as a regular Packet Boat between Glasgow and the western end of the Caledonian Canal at Fort William, by way of the Crinan Canal in Argyleshire. Mr Bell about the same time constructed the *Stirling Castle* Steam Boat, and employed her on the River Forth, between Leith and Stirling; he afterwards took her to Inverness, where she has been for two years plying between that town and Fort Augustus, going seven miles by the Caledonian Canal, and twenty three miles along Loch Ness."

Such praise and recognition of his contribution to the development of steam navigation from so distinguished a source must have gladdened Bell's heart. He recognised the value of the commendation from such an important source and was careful to reprint this part of the Select Committee's report in his appeal *To the Noblemen, Gentlemen and Freeholders of the County of* . . .

However not even the august authority of the House of Commons can, it seems, always guarantee total accuracy. The Committee's report gives an 1811 date for the *Comet* which is indeed contradicted by the appendix to the Report which gives the launch date as 1812. The positive assertion that Bell went out to America to assist Fulton is, from every possible source of information, not least Bell's own total silence on the subject, almost certainly wrong.

James Cleland's balanced judgement on the claims of the various contenders, which caused so much concern to Patrick Miller Jnr, is also worth serious consideration, based as it is on a wealth of personal and local knowledge.

> The ingenuity and indefatigable perseverance of Mr Miller of Dalswinton, will never be disputed, assisted as he was by Mr Symington, but like Rumsey, and Fitch of America, their exertions were of little or no avail to their country, having never been brought to any beneficial end, although supported by fortune—fortified by letters patent—and the command of the high political influence of Mr Miller's brother who had been formerly town clerk of Glasgow. In assigning the merits of the propelling system to Mr Miller or Mr Symington, it is difficult to account for their conduct in abandoning their joint schemes in 1802, and Mr Symington remaining inactive for ten years thereafter, while Mr Bell was going on with his experiments, for surely, if they could have brought their machinery to bear, the workings might have been transferred from the canal to the adjoining River Forth, where the banks would not have been injured. . .
>
> Without assuming for Mr Bell the merit of instructing Mr Fulton . . . or believing that he ever crossed the Atlantic, as has been erroneously stated elsewhere; there cannot be a doubt but that he was the first person in Europe who successfully impelled vessels by Steam, nor can it be doubted that by the effects of his ingenuity and perseverance, Steam Vessels have been sent from the Clyde, to almost every river in Europe; and Royal Mails conveyed from one port to another with a precision formerly impracticable. . .[17]

Cleland went on to discuss the lack of official and governmental recognition and the steps that were taken locally to support and honour Bell. He concluded:

> The success which has attended Mr Bell's exertions, is a brilliant proof of the diversity of talent. . . Who could have thought that such progress would have been made by an

unexperienced man, and that too, under the very eye of the great improver of the Steam Engine, who, though he had retired from active life in 1800, devoted his leisure hours to inventions, for 19 years thereafter.

Cleland's mention of the "great improver of the Steam Engine" is of course a reference to James Watt. Cleland was an objective although enthusiastic supporter of Bell but it should be noted that even he allows a note of surprise to creep in when he has to record the extraordinary fact that Bell had actually been the one to win the race. His preconception is clear, and not unreasonable. Surely it was indeed to have been expected that the great James Watt, in his years of retirement, with his leisure hours able to be devoted to inventions, should have applied himself to and solved the problem of steam power at sea. Perhaps some of the mixed views about Bell and his achievement do have their origin in the view that Watt, or if not Watt then someone of the stature of Thomas Telford, James Rennie or Brunel should have been the one to successfully produce marine steam power and not "an unexperienced man". Brunel's comment on this outcome, perhaps somewhat rueful in tone, is worth recollecting:

... he accomplished what others had failed in.[18]

The claim of priority in any scientific or technical development is notoriously difficult to sustain. Progress is often made by different workers independently and simultaneously. Scientific development is more accurately described as going forward on a wide front with a great many contributions going to the final result rather than as the popular idea of invention being the product of a solitary experimenter crying "Eureka". The Comet was so swiftly followed onto the Clyde by other steamers that one may feel that it had only been a matter of chance that Bell had got there first. However it is fair to point out that, due to Bell's problems over money, the Comet was a long time building. Her construction would hardly have been a secret, so that there was every opportunity for others to learn about it from visiting the yard or from local gossip.

As we have seen Bell had involved many other people, such as Robertson and Thomson and possibly Symington and Fulton, in discussion and experiment on the steam navigation problem. The possibility of steam navigation was, and had been for many years, a live, current issue and the subject of debate and discussion among interested circles. Despite this there had been a long gap in time and a very obvious hiatus in development between the laying up of the *Charlotte Dundas II* and the launch of the *Comet*. In these years, although Fulton was having public and well-reported success in America, in Britain and in Europe no one had, for various reasons, advanced matters beyond the point Symington had reached. There was no earlier, commercially successful, steamboat in Europe than the *Comet*. The great prize was still waiting to be won and Bell got there first.

Quite why Bell succeeded is a complex question which is not really adressed by Cleland's comment on the "diversity of talent". His early training and his seemingly deliberate and considered attempt to widen his experience had certainly introduced him to some of the skills that he would need to make a success of the technical aspects of steam power. Perhaps one can go further. Arguably the very diversity of this training and work experience—stonemason, millwright, shipbuilder, engineer, housebuilder, wright and architect—may indeed have given him only a fairly superficial grasp of these skills, but may have been in fact the essential quality that was needed. John Robertson, the engine builder for the *Comet* and himself a skilled engineer described Bell as:

> restless, unmethodical and by no means a good hand at machinery.[19]

If, as seems likely, *Comet*, when she ran ashore at Craignish, split apart where she had been lengthened, as a consequence of Bell's skimping and use of unsuitable timber, then one could suggest that his year's experience at Shaw and Hart's shipyard would seem to have been only of limited value to him. A case can certainly be argued that John Robertson was a better engine builder than Bell, John Wood a better shipbuilder than Bell, almost anyone a more prudent businessman than Bell.

But, having said all that, it was Bell and not Watt, Telford, or Rennie; Robertson, Wood or Napier; still less Miller, Taylor or Symington who claims the place in history and who saw the *Comet* into service. As there was no fundamental "invention" required, but rather the synthesis of already existing technologies and skills, one could argue that the "jack of all trades" with an open mind was what was required to succeed. To wide skills and an open mind can perhaps be added one other essential requirement, an all-consuming vision.

The comparison with the character and abilities of the American Robert Fulton is surely instructive. Fulton was, like Bell, a man of wide-ranging interests—portrait painter, patentee of flax spinning machines, inventor of marble sawing machinery. He was, again like Bell, a man of strong views and a man driven by a desire to succeed and a willingness to use any means to achieve his ends—Napoleonic France, Britain or his own United States government. Theirs was an age which gave birth to engineers of great brilliance and where engineers, perhaps for the first time, became major public figures, almost public heroes. The semi-heroic nature of the great engineers of the late eighteenth and early nineteenth century forms the sub-text to books like Samuel Smiles *Lives of the Engineers* with their promotion of the image of men like Watt, Rennie and George and Robert Stephenson. Neither Bell nor Fulton can be reasonably classed in the same league as a Watt, a Brunel or a Telford—but these two, perhaps somewhat eccentric, certainly unorthodox, unquestionably highly motivated but in many respects "unexperienced" men from the second or third division were the ones who gave steam navigation as a practical reality to the world. In so doing they refuted Watt's words to Bell:

> How many noblemen, gentlemen and engineers have
> puzzled their brains, and spent their thousands of pounds,
> and none of all these, nor yourself, have been able to bring
> the power of steam to a successful issue.[20]

Bell's achieving the goal which had eluded so many greater men is, and was to many, an unsettling thought. We have seen its capacity in this regard surfacing even in the writings of an informed and sympathetic contemporary like James Cleland.

However Bell's range of activities were, as we have seen, much wider than just building and running Europe's first practical steamboat. It is impossible, and probably pointless, to speculate about the source of the imagination and vision which led him into ploys like draining Loch Lomond, canalising the Clyde or cutting a ship canal across Kintyre.

He was evidently a physically and mentally restless man. A man forever seeking new areas in which to work and new plans to scheme over, an assiduous self-promoter, an unreliable businessman with a poor credit record and a spendthrift. Such a man could have been a deeply unattractive personality. Bell from all accounts was the opposite. He attracted a wide circle of loyal supporters and friends. These came even from among those whom he had let down financially, and from among the close-knit business community of Glasgow, a class whose business habits, civic pride, religious doctrines and group identity would make it unlikely for them to publicly overlook or forgive financial impropriety or irregularity. What was it in the man and his character that attracted such support and friendship?

His contemporary Gabriel MacLeod's description of him, as set down by Donald MacLeod in his *Nonogenerian's Reminiscences of Garelochside and Helensburgh*, perhaps gives some clues.

> As most of those who have seen his picture . . . may readily
> infer, he had a good-looking shrewd countenance—the face
> of an honest though somewhat positive Scot—sharp featured,

with high cheek bones and a clear grey eye. Latterly his face wore a shadow of care or anxiety in repose, which passed away when animated by conversation, and, especially if the topic was a favourite one, it melted into sunshine. His speech was quick and pointed, and he had a restless activity of manner in work. . .

Any man to succeed in life must have faith in himself and faith in his work. Bell had both. He was an enthusiast in science as applied to steam navigation, and like all enthusiasts was always ready to discuss his projects to any willing ear. Nothing, in fact, delighted him more than a good and appreciative listener, to whom he could pour out the whole story of his great discoveries and dreams of the future, and it was a great treat to spend an evening with him, listening to his varied experiences and hopes for the coming time.[21]

It is surely this quality of enthusiasm which distinguishes Bell from his rivals. His obituarist wrote of "his 'ruling passion' scheming" and of him as the "ingenious Mr Henry Bell", and these characterisations are just—but enthusiasm driven by faith in himself and in his plans and in his vision, his vision above all of the future of steam power, was surely dominant. It is inconceivable that Bell could ever have written, as James Taylor did, in these terms:

. . . a combination of unfortunate circumstances, over which I possessed no control, has prevented me from reaping any benefit from a discovery, which in the hands of a more pushing, or more lucky individual, might have raised him to opulence and distinction. . . [22]

Henry Bell's enthusiasm, vision and drive certainly failed to raise him to opulence, but he could, at the end of his days, look back and feel that he had made his mark on the world. He may have been plagued by financial worries but at least he avoided any regrets, like Taylor's, at not having taken the chance and put matters to the test. After his death a relative was to write of Taylor's

> habit of procrastination, combined with a certain cons-
> titutional indolence, which had crept on him with advancing
> age. . .[23]

This was not a charge that anyone could lay against Bell.

An interesting insight into Bell's character and beliefs comes, rather unexpectedly, in the Captain's Account Book of the *Comet*. This volume, which has been preserved by being handed down through a Helensburgh family associated with Bell, was presumably returned to Henry Bell, as ship-husband for the *Comet* by Captain Bain after the loss of the steamer in December 1820. The first forty pages of this book are in one hand, presumably that of Bain's. However at the back of the account book, in a different hand, and written with a complete disregard for punctuation, grammar and spelling, is a 900 word long religious meditation, which from all the internal evidence is undoubtedly the original, and previously unpublished, work of Henry Bell.

> Com unto me all ye that labor and are heave laden hear goad
> calls upon all that are heave laden to com unto him—what
> is it to be hevey laden it is to have a burthen of sin and we
> cannot of ourselfs take of this burthen but if we come unto
> god he shal take it away fore he is able to save to the
> outermost all thous that come unto him therefor why will
> we not come unto him when he is abel to save us for there is
> not another in the wourld that is abel to save us but also sin
> is the thing that is most loved it is nuriched and sherished as
> if it could make us hapy but the hapenes of sin is but of shourt
> duration it is but of a moment and then the sting that is left
> is most indurabel when you go to a Dance all your mind is
> taken up for a week before and then alas wheat is it when it
> comes is it a thing that will bring hapnes or comfort or gay
> or peace to your death bed o no it will not calm your

langushen hours but alas it will rather add a pang yeas thear is not a sin which does not leave a pang to a reflecting mind you will say what a happy night we have had but alas what was it in a fine party or a fine asembley of pepal when there was purhaps every thing to dasel the mind and to turn the giddy head is this hapness o no it cannot how could it a number of pepel meet togther for merment is there anething like hapness here no there cannot where then is pepal that think there is hapness in these things there is littel happnes for a son as all the companey is goan the the langushen hours begin and they canot rest if it be a weife that is faund of companey her own husbands society becomes irksome o can there be hapnes hear no there cannot if it be the husband his own family becomes irksom he goes out from the bosom of his famley to seak what he calls hapness what is it it is purhaps a larg bool of toady and a wheen merrey fellows but alas if this be persisted in o what hapnes it will lead to it will lead to Destrucion and o when that man perhapes comes to a death bead it will stare him in the face and he will exclame o that i could begin my life over agin i would not seak hapneas with druncards but with god for in him alone is true hapnes and though he has kindly aloud us six days of the weak to do our ane work yeat we are verey weered if we have one to serve him we get tired with his serves very soon when come home from the hous we apere to think the sabth over we are restles if we have not compney we toss our selfs too and fro until we ether march into the feldes or get into idel conversation with our nabors we prefare that to comuenin with god but the tim will come as is in the above that we would wish o that i had but my life to live over agen or perhapes o that i had never been born o how lamentabel moust it be to be in this stat when our acquentens our frinds look angurly at us it gives us a pang that we cannot expres but how littel we are afraid to incur the displesure of god it gives us as it were no unesens but o when that awful moment comes then alas what must be the tourtor we will feel about to enter upon eternity o eternity that awful wourd everlasting punishment and will we not cry unto god to take us from this state this aful state for he is able and willing to take us unto himself if we would but coem unto him and live.[24]

257

This remarkable effusion, is perhaps somewhat better read in an edited version with spelling corrected, some punctuation added and some minor textual additions (indicated by square brackets) made to assist understanding.

Come unto me all ye that labour and are heavy laden. Here God calls upon all that are heavy laden to come unto him. What is it to be heavy laden? It is to have a burden of sin and we cannot of ourselves take off this burden, but if we come unto God he shall take it away, for he is able to save to the uttermost all those that come unto him. Therefore why will we not come unto him when he is able to save us, for there is not another in the world that is able to save us. But also sin is the thing that is most loved, it is nourished and cherished as if it could make us happy, but the happiness of sin is but of short duration, it is but of a moment and then the sting that is left is most unendurable.

When you go to a dance all your mind is taken up for a week before and then, alas, what is it when it comes? Is it a thing that will bring happiness or comfort or joy or peace to your death bed? O no! It will not calm your languishing hours but alas it will rather add a pang. Yes, there is not a sin which does not leave a pang to a reflecting mind. You will say, "what a happy night we have had", but alas what was it in a fine party or a fine assembly of people when there was perhaps every thing to dazzle the mind and to turn the giddy head. Is this happiness? Oh! no it cannot [be], how could it? A number of people meet togther for merriment; is there anything like happiness here? No there cannot [be]. Where then is people that think there is happiness in these things? There is little happiness for as soon as all the company is gone then the languishing hours begin and they cannot rest. If it be a wife that is fond of company her own husband's society becomes irksome. Oh! can there be happiness here? No there cannot. If it be the husband his own family becomes irksome, he goes out from the bosom of his family to seek what he calls hapiness. What is it? It is perhaps a large bowl of toddy and a wheen merry fellows, but alas, if this be persisted in, Oh! what

[un]happiness it will lead to. It will lead to destruction and Oh! when that man perhaps comes to a death bed it will stare him in the face and he will exclaim Oh! that I could begin my life over again, I would not seek happiness with drunkards but with God, for in him alone is true happiness.

And though he has kindly allowed us six days of the week to do our own work yet we are very wearied if we have one to serve him. We get tired with his service very soon. When [we] come home from the House [of God?] we appear to think the Sabbath over. We are restless if we have not company, we toss ourselves to and fro until we either march into the fields or get into idle conversation with our neighbours. We prefer that to communing with God, but the time will come as is in the above that we would wish "Oh! that I had but my life to live over again", or perhaps, "Oh! that I had never been born". Oh! how lamentable must it be, to be in this state.

When our acquaintances, our friends look angrily at us, it gives us a pang that we cannot express. But how little we are afraid to incur the displeasure of God. It gives us, as it were, no uneasiness. But Oh! when that awful moment comes then alas what must be the torture we will feel about to enter upon. Eternity, Oh! eternity, that awful word. Everlasting punishment and will we not cry unto God to take us from this state, this awful state? For he is able and willing to take us unto himself if we would but come unto him and live.

One would not wish to claim too much for this piece. It will be recollected that Margaret Bell was an enthusiastic churchwoman but little has been recorded previously of Henry's religious views. This very unstructured meditation on the search for happiness and the transitory nature of worldly pleasures hardly puts Bell in the first rank of theological thinkers but it is surely of interest as an indication of a thoughtful and religious disposition, or in the words of the meditation "a reflecting mind". It is also surely very characteristic of Bell in its colourful and direct imagery— the dance metaphor, the description of the reaction to the Sabbath:

> We are restless if we have not company, we toss ourselves to
> and fro until we either march into the fields or get into idle
> conversation with our neighbours. . .

the reference to the attractions of a "bowl of toddy and a wheen
merry fellows". One is reminded of his letter to Seaforth on the
subject of the West Highland steamer service with its outpouring
of ideas and images and the sharp reflections on the priorities of
the Highland landowners.

One interesting reflection on the perception of Bell's
contribution to marine steam power comes from the United
States. It would seem that an annual report was sent to Bell on
the progress of steamboats on the River Mississippi. One of these
reports was published in the *Glasgow Courier* in 1823 and
consisted of a list of ships plying on the Mississippi and its
tributaries together with the reflection:

> . . . what a great advantage it is to the public at large since
> you first introduced them into your rivers and lakes. . .[25]

The fact that an American correspondent, unfortunately
unnamed, thought it appropriate to write to Bell in these terms
and to provide a list of steamships indicates a contemporary
international appreciation of Bell's role and claim to recognition.
A significance all the greater when it is recollected that the
American Fulton had strenuously argued that he and he alone
was responsible for the development of the steamship. A similar
sort of recognition from the wider world can be seen, again in
1823, in the decision of the Mersey and Clyde Steam Navigation
Company to give Bell's name to a new 200 ton, 60 horse power
steamer they had built for their Glasgow to Liverpool route. The
Henry Bell joined their other two steamers on this trade which
were named after James Watt and William Huskisson, President
of the Board of Trade and Bell's inclusion in this distinguished
group is surely significant of a widespread recognition of his
claims.

In 1912 the centenary of the inaugural voyage of the *Comet* was marked with large scale celebrations organised by the City of Glasgow and the county and burgh councils around the Clyde, with the co-operation of the other public bodies concerned with the Clyde, the trade and professional bodies and the Glasgow University and the Royal Technical College. A major exhibition was held at Glasgow's Kelvingrove Museum, funds were to be raised to purchase a ship for the use of students of navigation and marine engineering at the Royal Technical College, firework displays were held, the Bell monuments throughout the country were decorated and a public holiday was proclaimed for Saturday 31st August. Souvenir editions of newspapers were produced and commemorative publications were issued. On the 30th August Glasgow City Council gave a luncheon in the City Chambers at which T McKinnon Wood, MP, Secretary of State for Scotland proposed the memory of Henry Bell.

The highlight of the celebrations was the cruise, for:

A specially invited party of representatives of public
bodies . . . on board the R.M.S. *Columba* . . .[26]

down river from the Broomielaw, in the track of the *Comet* to the Tail of the Bank, off Greenock. The distinguished guests in Messrs MacBrayne's elegant steamer *Columba* were followed in the Glasgow Corporation sewage sludge boat TSS *Shieldhall* by:

. . . representatives of the shipbuilding industry, consisting
of Shipwrights, Engineers, Boilermakers and other trades.. .

and by members of the Navy League in the specially chartered paddle steamer *Queen Empress*. On their way down river the voyagers would have been able to see in the twenty nine ship-building yards that they passed no fewer than one hundred and seventy five vessels under construction or fitting out. This demonstration of the shipbuilding and engineering pre-eminence of the Clyde encompassed ships of every size and type. A bucket dredger in Lobnitz & Company's yard, warships ranging in size from a torpedo boat to a battleship, the *Aquitania* in John Brown's of Clydebank, which when launched in April 1913 would be, at 45,647 tons and 901 feet in length the largest vessel afloat. There is a century of truly startling progress from the 25 ton *Comet*, small enough and light enough to be pushed off a Clyde sandbank by her crew, to the beauty of the majestic *Aquitania*, designed to carry 3200 passengers and nearly a thousand crew across the Atlantic at 24 knots. But the message that the packed Clyde shipyards had to tell the centenary excursionists was clear: all this had come from that first tentative voyage "by the power of Wind, Air and Steam."

Bell's vision of the role of steam power at sea was perhaps nowhere more clearly vindicated than at the Tail of the Bank, the roadstead off Greenock, the town that had been the destination for the *Comet*'s first commercial voyage. Here, and off Dunoon and, appropriately enough, off his own adopted town of Helensburgh, was arrayed an impressive display of British shipping. Yachts, tugs, dredgers and the merchant vessels of twenty Scottish-based shipping lines assembled to honour Bell's memory. They were complemented by a powerful battle squadron of the Royal Navy's Home Fleet. Five battleships, two cruisers and four destroyers under the command of Vice Admiral Sir John Jellicoe provided a tribute to the man and his work which was a far cry from the dismissive comment that the First Lord of the Admiralty, Lord Melville, had written on Bell's letter back in 1813:

> This is not by many the first proposal of a Vessel to be driven by steam, and in this as in all other inventions, the request is that John Bull should pay the piper. Steam vessels may make progress in Canals and still water but to contend with the Atlantic Ocean and other seas—the projector may as well attempt to "bottle off" these great National Waters. [27]

John Bull never did quite "pay the piper" as far as Bell was concerned but, however belatedly, "the ingenious Mr Henry Bell" did achieve the recognition and thanks of his country.

NOTES

CHAPTER 1

1) Glasgow Herald 10th August 1812
2) Quoted in Robert Chambers: Lives of Eminent Scotsmen 1872 p118
3) Glasgow Herald 14th August 1812
4) Peter Mackenzie: Old Reminiscences of Glasgow and the West of Scotland Vol 2 1890 pp 273ff
5) Andrew McGeorge: Old Glasgow, the place and the people 3rd ed 1888 p245
6) James Pagan: Glasgow Past and Present Vol.2 1851 p238
7) Plans of *Comet* Glasgow University Archives, Napier Collection DC90/2/4
8) Quoted in Glasgow Herald 24/7/1912
9) John Buchanan's notes of conversation with John Robertson, National Library of Scotland MSS 2675
10) Memorandum by William Mackenzie, dated 26th January 1820, transcribed in 1898 by David Bell from the original document then in possession of Messrs Robert Napier & Co, transcript in Museum of Transport, Glasgow
11) Memoir of David Napier by himself, Glasgow University Archives, Napier Collection DC90/2/14
12) Promissory Note by Henry Bell to David Napier, Glasgow University Archives, Napier Collection DC90/2/14
13) Decree of Horning 30th March 1813 Strathkelvin District Libraries, Archives Collection T20/12
14) John Buchanan's notes of conversation with John Robertson op cit
15) Customs House Records: Scottish Ports: 1820 Public Record Office BT107 403
16) John Buchanan's notes of conversation with John Robertson op cit
17) John Buchanan's notes of conversation with John Robertson op cit
18) Henry Bell: Observations on the Utility of Applying Steam Engines to Vessels etc, Glasgow 1813
19) John Buchanan's notes of conversation with John Robertson op cit
20) Henry Bell: op cit

21) Henry Bell: op cit
22) Manuscript list of Steam Boats built at Port Glasgow by Mr Wood since the year 1812, Glasgow University Archives, Napier Collection DC90/2/4
23) James Cleland: Annals of Glasgow Vol 2 1816 pp394/5
24) Peter Mackenzie: op cit
25) William Harriston: The steam-boat traveller's remembrancer. . . Glasgow 1824, Reprinted The Molendinar Press, Glasgow 197?
26) Edward Morris: The Life of Henry Bell 1844 pp170/173
27) "Britannicus": To the Genius of Helensburgh, quoted in Edward Morris op cit pp147/8
28) Letter by James Cook to James Cleland dated 4th April 1825, quoted in James Cleland: Historical Account of the Steam Engine and its application in propelling vessels . . . Glasgow 1829 pp51/2
29) Glasgow Chronicle, quoted in Glasgow Herald 19th November 1830
30) John Buchanan's notes of conversation with John Robertson op cit
31) Edward Morris: op cit pp98 & 105

CHAPTER 2

1) Edward Morris : The Life of Henry Bell 1843 p15
2) op cit p16
3) McUre :History of Glasgow 1830 p247
4) New Statistical Account of Linlithgowshire: Parish of Torphichen 1842
5) Edward Morris: op cit p81
6) op cit p25
7) MSS letter by Bell to Dr Archibald Wright 16th August 1820, Strathkelvin District Libraries Archives Ref T20/11/1
8) Edward Morris: op cit p16
9) op cit p16
10) op cit p17

CHAPTER 3

1) Jones' Glasgow Directory, Glasgow 1789
2) Burgesses & Guild Brethern of Glasgow 1751–1846, Edinburgh Scottish Record Society 1935
3) The Glasgow Directory 1801–2, Glasgow 1801
4) The Glasgow Directory 1804, Glasgow 1804
5) Edward Morris: The Life of Henry Bell 1843 pp 148/9
6) Glasgow University Archives Scotts Shipbuilding & Engineering Coy Outgoing letter book 1806 GD319/11/1/2
7) Dumbarton Herald 15th May 1856
8) James Barr: Balloch and around, Transcript in Dumbarton Public Library of articles first appearing in Dumbarton Herald 1892-93

9)   Edward Morris: op cit 1843 p 63
10)   Public Record Office: Customs House Transcripts and Transactions
      Series 1 Scottish Ports 1820
11)   John Buchanan's notes of conversation with John Robertson, National
      Library of Scotland MSS 2675
12)   Scottish Record Office, General Register of Sasines: Dumbartonshire
      No 2060 16th May 1810
13)   Scottish Record Office, General Register of Sasines: Dumbartonshire
      No 3221 2nd December 1820
14)   Scottish Record Office, Heritors' Records, Parish of Carluke 1761–1806
      HR179/1
15)   Edward Morris: op cit 1843 p17
16)   Edward Morris: op cit 1843 p29
17)   James Cleland: Annals of Glasgow. . . Glasgow 1816 Vol 2 p514
18)   Memoir of David Napier by himself, Glasgow University Archives,
      Napier Collection. DC90/2/4
19)   Fifth Report of the Select Committee appointed to inquire into the state
      of the Roads from London to Holyhead. . . PP 1822 VI 417
20)   Chambers Edinburgh Journal 5th January 1839
21)   Henry Bell: To the Committee and other Subscribers interested in
      Supplying the City of Glasgow with Water Glasgow, 1806 Strathclyde
      Regional Archives TD200/85C
22)   Scottish Record Office, General Register of Sasines, Glasgow No 10332
      21st December 1815
23)   John Buchanan's notes of conversation with John Robertson op cit
24)   Thomas Thomson: A biographical dictionary of Eminent Scotsmen. . .
      London 1870 Vol 1 p118

CHAPTER 4

1)   Supplement to the Fourth, Fifth, and Sixth Editions of The
      Encyclopedia Britannica . . . Edinburgh 1824 Vol 6 p537
2)   op cit p538
3)   James Taylor to William Huskisson MP, President of the Board of Trade
      19th May 1825, Scottish Record Office GD51/1/466
4)   William Symington: Letter in Edinburgh Courant 12th September 1825
5)   Taylor to Huskisson op cit
6)   Scots Magazine November 1788
7)   Caledonian Mercury 20th May 1790
8)   Quoted in A Ian Bowman: Symington and the *Charlotte Dundas*, Falkirk
      1981 p16
9)   James Taylor to William Huskisson MP, President of the Board of Trade
      19th May 1825, Scottish Record Office GD51/1/466
10)   Henry Bell: Letter in Caledonian Mercury 9th October 1816
11)   Patrick Miller Jnr: Memoir regarding Symington and Bell's pretensions
      to be considered the original Inventors of Steam Navigation . . .

Edinburgh, New Philosophical Journal April/October 1827.

12) Patrick Miller Jnr to Viscount Melville, First Lord of the Admiralty 28th April 1823, National Library of Scotland MSS 1055 f14

13) Patrick Miller Jnr to Archibald Constable 3rd May 1825, National Library of Scotland MSS 356 ff22–23

14) James Cleland: Historical Account of the Steam Engine . . . Glasgow 1825

15) James Cleland to Alexander McGregor 12th May 1825, National Library of Scotland MSS 356

16) William Symington: Letter in Edinburgh Courant 12th September 1825

17) Patrick Miller Jnr to William Huskisson 2nd August 1825, Scottish Record Office GD51/1/466

18) Defences for Henry Bell Jnr of the Baths, Helensburgh, Proprietor of "The *Comet*" Steam Boat against William Symington, Engineer, formerly at Kinnaird, now at Falkirk, Scottish Record Office CS238 5/20/83

19) T Thomson: Biographical Dictionary of Famous Scotsmen 1875

CHAPTER 5

1) Glasgow Journal 11th January 1776

2) Helensburgh Town Council Minutes, Dumbarton District Libraries Archive Collection.

3) Donald MacLeod: A Nonogenarian's Reminiscences of Garelochside and Helensburgh . . . Helensburgh 1883 p151

4) Donald MacLeod: op cit p150

5) Scottish Record Office, General Register of Sasines, Dumbartonshire No 1680 23rd July 1806

6) Dumbarton Herald 15th May 1856

7) Scottish Record Office, Register of Deeds RD5/191 p437 recorded 28.11.1820

8) Scottish Record Office, General Register of Sasines, Dumbartonshire No 2060 16th May 1810

9) Scottish Record Office, General Register of Sasines, Dumbartonshire No 106 19th September 1831

10) Glasgow Courier 2nd December 1820

11) Greenock Advertiser 15th June 1821

12) Scottish Record Office, Dumbarton Sheriff Court Inventory of Personal Estate of Henry Bell SC65/34/2

13) Glasgow Courier 2nd December 1820

14) Glasgow Courier 22nd December 1820

15) John Galt: The Steam Boat in The Provost and other tales, London, MacLaren & Co, nd p177

16) Scottish Record Office, Register of Deeds RD5/191 p437 recorded 28.11.1820

17) Glasgow Courier 30th August 1823

18) Scottish Record Office, General Register of Sasines, Dumbartonshire No 2736 1st August 1816
19) Scottish Record Office, General Register of Sasines, Dumbartonshire No 939 16th June 1828
20) The Topographical, Statistical and Historical Gazetteer of Scotland, Glasgow 1842 Vol 1 p771
21) Dumbarton Herald 29th September 1853
22) Dumbarton Herald 15th May 1856
23) Harriet Beecher Stowe: Sunny Memories, London 1889 p215
24) Dumbarton Herald 7th April 1853
25) Dumbarton Herald 15th May 1856
26) Glasgow Herald 5th May 1856
27) Dumbarton Herald 15th May 1856

CHAPTER 6

1) Bell to First Lord of the Admiralty 2nd April 1813, Public Record Office, Admiralty Promiscuous Letters 1813–14 ADM1 4383 fB51
2) William Bain: Remarks on the Progress of Steam Navigation Blackwood's Edinburgh Magazine November 1825
3) Edward Morris: The Life of Henry Bell, Glasgow 1843 pp17/18
4) Public Record Office, Admiralty Board Minutes January–June & July–December 1800 ADM3 124 & 125
5) Public Record Office, Admiralty Board Minutes January–June & July–December 1803 ADM3 148 & 149
6) Public Record Office, Admiralty Secretary's Indexes & Compilations Series ADM 12
7) Public Record Office, Admiralty Promiscuous Letters 1803–4 ADM1 4378
8) Public Record Office, Admiralty Promiscuous Letters 1813–14 ADM1 4383
9) Public Record Office, Admiralty Secretary's Indexes & Compilations Series ADM 12
10) Letter by Henry Bell, Caledonian Mercury 9th October 1816
11) Henry Bell: Observations on the Utility of Applying Steam Engines to Vessels etc 1813
12) Henry Bell: "To the Noblemen, Gentlemen and Freeholders of the County of . . ." 1826
13) William Bain: Remarks on the Progress of Steam Navigation, Glasgow Mechanics Magazine 12th November 1825
14) H W Dickinson & R Jenkins: James Watt and the Steam Engine, Ashbourne 1981 p324
15) H W Dickinson: Robert Fulton: Engineer and Artist—his life and works, London 1913 pp181/185
16) Samuel Smiles: Lives of the Engineers, John Rennie, London 1904 p392
17) quoted in H W Dickinson op cit pp179/80

18) James Taylor to William Huskisson, MP President of the Board of Trade 19th May 1825 Scottish Record Office GD51/1/466
19) Letter by Henry Bell, Caledonian Mercury 9th October 1816
20) Edward Morris: op cit pp31/32
21) Edward Morris: op cit pp74/75
22) Samuel Smiles: op cit p392
23) W S Harvey & G Downs-Rose: William Symington; inventor and engine-builder, London 1980
24) Patrick Miller Jnr to 2nd Viscount Melville, First Lord of the Admiralty 28th April 1823 National Library of Scotland MSS 1055 f14
25) Cynthia Owen Philip: Robert Fulton; a biography, New York 1985 pp 262/265
26) Quoted in William Charles Maugham: Annals of Garelochside Paisley 1897 p46

CHAPTER 7

1) William Bain: Remarks on the Progress of Steam Navigation, Blackwood's Edinburgh Magazine November 1825
2) John Thomson: Account of a Series of Experiments, made for the Purpose of Ascertaining the Best Mode of Constructing Vessels with Machinery, to be Wrought on Navigable Rivers by the Power of Steam. Including a Review of a Pamphlet, lately Published, intitled, 'Observations on the Utility of Applying Steam Engines to Vessels, etc By Henry Bell' Glasgow 1813
3) Henry Bell: Observations on the Utility of Applying Steam Engines to Vessels etc, Glasgow 1813 pp13/14
4) John Thomson: op cit p12
5) Edward Morris : The Life of Henry Bell, Glasgow 1843 p25
6) Edward Morris: op cit p28
7) Letter by James Taylor to William Huskisson MP, President of the Board of Trade, dated 19th May 1825 Scottish Record Office GD51/1/466
8) John Thomson: op cit p8
9) William C Maugham: Annals of Garelochside. . . Paisley 1897 p40/41
10) John Thomson: op cit p10
11) John Buchanan's notes of conversation with John Robertson, National Library of Scotland MSS 2675
12) John Thomson: op cit p4/5
13) John Thomson: op cit p10
14) John Thomson: op cit p12
15) John Thomson: op cit p17
16) Letter by James Taylor to William Huskisson MP op cit
17) Edward Morris: op cit p56
18) Edward Morris: op cit p57
19) Donald MacLeod: A Nonogenarian's Reminiscences of Garelochside

and Helensburgh, Helensburgh 1883 pp153/154

20) Edward Morris: opcit p25
21) Undated quotation from Glasgow Herald c1912, Science Museum, Rhys Jenkins Collection of Newspaper Cuttings 77c
22) Glasgow Herald 12th February 1813
23) Glasgow Herald 2nd July 1813
24) Glasgow Courier 2nd April 1814
25) Glasgow Courier 24th May 1814
26) Glasgow Courier 24th November 1814
27) Glasgow Courier 28th September 1815
28) James Cleland: Historical account of the Steam Engine and its Application in Propelling Vessels. . . Glasgow 1829 p59
29) Glasgow Mechanics Magazine 5th February 1825, Letter by Henry Bell
30) Edinburgh Evening Courant 24th May 1813
31) John Thomson: op cit p19
32) Glasgow Courier 8th September 1814
33) Glasgow Courier 9th September 1815
34) Scots Magazine February 1817 pp106/108, Letter by Henry Bell "Plan of Communication by Steam Boats between Leith and Greenock"
35) Glasgow Herald 28th June 1813
36) Glasgow Courier 13th September 1814
37) Annual Register 1817 Chronicle p101

CHAPTER 8

1) Glasgow Courier 13th May 1815
2) James Cleland: Annals of Glasgow Vol 2, Glasgow 1816 p396
3) Glasgow Mechanics Magazine, Letter by Henry Bell 5th February 1825
4) Letter by Henry Bell to Sir Hugh Innes 23rd December 1819, Scottish Record Office GD46/17/53
5) Glasgow Courier 18th November 1815
6) Scots Magazine March 1816 pp164–167 J C Delametherie "On Steam Vessels: with a narrative of a Voyage performed by one from Glasgow to London"
7) Glasgow Courier 19th September 1815
8) Glasgow Courier 30th March 1820
9) Edward Morris: The Life of Henry Bell, Glasgow, 1843 p154
10) Public Record Office: Customs House Transcripts and Transactions Series 1, Scottish Ports 1820 BT 107 403
11) Edward Morris op cit p154/5
12) Glasgow Courier 31st August 1819
13) The Scotsman 17th August 1822
14) Letter by Henry Bell to Sir Hugh Innes 23rd December 1819, Scottish Record Office GD46/17/53
15) Letter by Sir Hugh Innes to J A Stewart MacKenzie of Seaforth 2nd May 1820, Scottish Record Office GD46/17/56

16) Glasgow Chronicle 28th November 1820

17) Letter by J A Stewart MacKenzie of Seaforth to Henry Bell 13th July 1820, Scottish Record Office. GD46/1/52

18) Letter by Henry Bell to J A Stewart MacKenzie of Seaforth 16th August 1820, Scottish Record Office. GD46/17/57

19) Seaforth Papers, List of Subscribers to Hebridean steamboat, dated London 4th April 1821, Scottish Record Office GD46/17/57

20) Letter by Sir Hugh Innes to J A Stewart MacKenzie of Seaforth 15th June 1821, Scottish Record Office GD46/17/57

21) Glasgow Courier 6th June 1820

22) Greenock Advertiser 19th December 1820

23) Public Record Office: Customs House Transcripts and Transactions Series 1 Scottish Ports 1820 BT 107 403

24) Comet Steam Boat Company: Constitutional Resolutions adopted at meeting of 30th September 1820, Strathkelvin District Libraries Archives T20/12/1

25) *Comet* Steam Boat Accounts 1820 Private Collection

26) Comet Steam Boat Company: Constitutional Resolutions op cit

27) Letter by Henry Bell to Dr Archibald Wright 16th August 1820, Strathkelvin District Libraries Archives T20/11/1

28) Letter by Henry Bell to Dr Archibald Wright [23rd August 1820?] Strathkelvin District Libraries Archives T20/11/1

29) Greenock Advertiser 19th December 1820

30) *Comet* Steam Boat Accounts 1820, Private Collection

31) Donald MacLeod: A nonogenarian's reminiscences of Helensburgh and Garelochside . . . Helensburgh 1883 p158

32) John Buchanan's notes of conversation with John Robertson, National Library of Scotland MSS 2675

33) Edward Morris: op cit p156

34) Edward Morris: op cit p155

35) Public Record Office: Customs House Transcripts and Transactions Series 1, North British Ports 1822 BT 107 405

36) Supplement to the Fourth, Fifth and Sixth Editions of the Encyclopedia Britannica . . . Volume Sixth "Steam Navigation" p545 Edinburgh 1824

37) Edward Morris: op cit p156

38) Public Record Office: Customs House Transcripts and Transactions Series 1, Northern Ports (North Britain) 1814 BT 107 113

39) Glasgow Courier 27th July 1819

40) *Comet* Steam Boat Accounts 1820, Private Collection

41) Glasgow Courier 11th April 1820

42) Joseph Mitchell: Reminiscences of my Life in the Highlands Vol 1, Newton Abbot 1971 p223/4

43) Letter by Sir Hugh Innes to J A Stewart MacKenzie of Seaforth 2nd May 1820, Scottish Record Office GD46/17/56

44) Public Record Office: Customs House Transcripts and Transactions Series 1, North British Ports 1824 BT 107 408

45) Letter from Alex Anderson & Coy Inverness to J A Stewart MacKenzie of Seaforth 4th July 1820, Scottish Record Office GD46/17/54
46) Glasgow Courier 18th November 1823
47) Joseph Mitchell: op cit p131
48) 20th Report of the Caledonian Canal Commissioners 1823 PP VII 412
49) Letter from Dugald Ferguson Jnr & Coy, Greenock to J A Stewart MacKenzie of Seaforth 16th December 1820, Scottish Record Office GD47/17/55
50) Glasgow Courier 18th September 1823
51) *Comet* Steam Boat Accounts 1820, Private Collection
52) Glasgow Mechanics Magazine 29th October 1825 Henry Bell: On the Late Lamentable Accident at Greenock, and Plans for Preventing the Like Occurence in Future.
53) Edward Morris: op cit p157
54) Quoted in Scots Magazine November 1825 p622
55) Quoted in Scots Magazine November 1825 p628
56) Quoted in Scots Magazine November 1825 p627
57) Glasgow Courier 13th July 1820
58) Public Record Office: Customs House Transcripts and Transactions Series 1, North British Ports 1822 BT 107 405
59) Public Record Office: Customs House Transcripts and Transactions Series 1, Scottish Ports 1821 BT 107 404
60) Glasgow Courier 22nd January 1828
61) Jean Lindsay: The Canals of Scotland, Newton Abbot 1968 p159
62) 20th Report of the Caledonian Canal Commissioners 1823 PP VII 412
63) 19th Report of the Caledonian Canal Commisioners—quoted in Glasgow Courier 18th July 1822
64) 20th Report of the Caledonian Canal Commissioners 1823 PP VII 412
65) Inverness Courier 26th August 1824—quoted in E Mairi MacArthur: Iona, the living memory of a crofting community 1750–1914 Edinburgh 1990

CHAPTER 9

1) National Library of Scotland, Paton Collection Acc 3219 f162
2) William Maugham: Annals of Garelochside, Paisley 1897 p48
3) op cit p48
4) Letter by Henry Bell to J A Stewart MacKenzie of Seaforth 23rd December 1819, Scottish Record Office GD46/17/53
5) Donald Macleod: A nonogenarian's reminiscences of Garelochside and Helensburgh . . . Helensburgh 1883 p154
6) Edward Morris: The Life of Henry Bell, Glasgow 1843 p120
7) op cit p63
8) Henry Bell: Letter to the Lord Provost of Glasgow . . . containing the outline of a Plan for the Improvement of the Harbour . . . Glasgow 1820
9) Edward Morris: op cit p12/13

10) View of the Merchants House of Glasgow, Glasgow 1866 p341
11) Edward Morris: op cit pp.90/91
12) op cit p95
13) op cit p96
14) op cit p120
15) Scottish Record Office, General Register of Sasines, Dumbartonshire 2736 1st August 1816
16) Glasgow Courier 26th January 1828
17) Edward Morris: op cit p134
18) Glasgow Chronicle quoted in Glasgow Herald 19th November 1830
19) Letter by Henry Bell to Walter Logan dated 11th January 1827, Dumbarton District Libraries Archives Collection
20) Edward Morris: op cit p131
21) op cit p130
22) op cit p133
23) op cit p133
24) Glasgow Chronicle quoted in Glasgow Herald 19th November 1830
25) Donald MacLeod op cit p154
26) Fowler's Commercial Directory of the Lower Ward of Renfrewshire for 1836–37 pp207/208
27) Inventory of Personal Estate of Henry Bell, Dumbarton Sheriff Court Records, Scottish Record Office SC65/34/2
28) Inventory of Personal Estate of Margaret Young or Bell, Dumbarton Sheriff Court Records, Scottish Record Office SC65/34/8

CHAPTER 10

1) John Thomson: Account of a Series of Experiments, made for the Purpose of Ascertaining the Best Mode of Constructing Vessels with Machinery, to be Wrought on Navigable Rivers by the Power of Steam. Including a Review of a Pamphlet, lately Published, intitled, 'Observations on the Utility of Applying Steam Engines to Vessels etc By Henry Bell' Glasgow 1813 p19
2) James Taylor to William Huskisson MP, President of the Board of Trade 19th May 1825, Scottish Record Office GD51/1/466/1
3) Patrick Miller Jnr to Archibald Constable 3rd May 1825, National Library of Scotland MSS 356 ff22–23
4) Patrick Miller Jnr to Viscount Melville, First Lord of the Admiralty 28th April 1823, National Library of Scotland MSS 1055 f14
5) Patrick Miller Jnr to Archibald Constable op cit
6) Patrick Miller Jnr: Memoir regarding Symington & Bell's pretensions to be considered the original Inventors of Steam Navigation . . . in The Edinburgh New Philosophical Journal April–October 1827 p88
7) William Symington: Letter in Edinburgh Courant 12th September 1825

8)   Patrick Miller Jnr: Memoir . . . op cit p90

9)   Sir Francis Drake: Dispatch to Sir Francis Walshingham 17th May 1587

10) James Taylor to William Huskisson op cit

11) James Taylor to William Huskisson op cit

12) James Taylor to William Huskisson op cit

13) William Symington op cit

14) Edward Morris: The Life of Henry Bell, Glasgow 1847 p105

15) Dumbarton Herald 7th April 1853

16) Fifth Report of the Select Committee appointed to inquire into the state of the Roads from London to Holyhead . . . PP 1822 VI 417

17) James Cleland: Historical account of the Steam Engine and its application in propelling vessels . . . Glasgow 1829 pp58–60

18) Edward Morris op cit p105

19) John Buchanan's notes of conversation with John Robertson, National Library of Scotland MSS 2675

20) Quoted in Robert Chambers: Lives of Eminent Scotsmen 1872 p118

21) Donald MacLeod: A Nonogenarian's Reminiscences of Garelochside and Helensburgh . . . Helensburgh 1883 pp149–152

22) James Taylor to William Huskisson MP, President of the Board of Trade August 20th 1825, Scottish Record Office GD51/1/466/2

23) Anon: A brief account of the rise an early progress of Steam Navigation: intended to demonstrate that it originated in the suggestions and experiments of the late Mr James Taylor of Cumnock, in connexion with the late Mr Miller of Dalswinton, Ayr 1844 [I am indebted to Gerard Quail for this reference]

24) *Comet* Steam Boat Accounts 1820 Private Collection

25) Glasgow Courier 30th August 1823

26) Celebration of Centenary of launch of steamer "*Comet*" built for Henry Bell, Official Programme, Glasgow [1912] p46

27) Bell to First Lord of the Admiralty 2nd April 1813 Public Record Office, Admiralty Promiscuous Letters 1813–14 ADM1 4383 ff351

SOURCES

As mentioned in the text, the only substantial attempt at a biography of Heny Bell was the work of Edward Morris, published in Glasgow in 1844.

Bell's engineering contemporaries have fared better and the following books, while shedding only limited light on Bell and his activities, are valuable for background information on Bell and his times.

Bowman, A Ian   *Symington and the Charlotte Dundas* Falkirk 1981
Dickinson, H W   *Robert Fulton, Engineer and Artist—his life and works* London 1913
Dickinson, H W & Jenkins, R
　　　　　　James Watt and the Steam Engine Ashbourne 1981
Harvey, W S & Downs-Rose, G
　　　　　　William Symington: inventor and engine-builder London 1980
Philip, Cynthia Owen
　　　　　　Robert Fulton: a biography New York 1985
Smiles, Samuel   *Lives of the Engineers John Rennie* London 1904
Spratt, H Philip   *The birth of the steamboat* London 1958

The environment in which Bell lived and worked, Scotland in the early nineteenth century, has a vast literature which is easily traceable. The following works, mainly of local history, have been drawn on for background and for specific information.

Cleland, James   *Annals of Glasgow* 2 vols Glasgow 1816
Lindsay, Jean   *The Canals of Scotland* Newton Abbot 1968
MacGeorge, Andrew
               Old Glasgow, the place and the people 3rd ed
               Glasgow 1888
Mackenzie, Peter *Old Reminiscences of Glasgow and the West of*
               Scotland 1890
MacLeod, Donald *A Nonogenerian's reminiscences of Garelochside*
               and Helensburgh . . . Helensburgh 1883
Maughan, William C
               Annals of Garelochside Paisley 1897
Mitchell, Joseph *Reminiscences of my life in the Highlands* 2 vols
               Newton Abbot 1971
Pagan, James   *Glasgow past and present* Glasgow 1851

BIBLIOGRAPHY

The archival sources for Henry Bell, on which this biography has been largely based, are detailed in the source notes for each chapter. However particular attention should be drawn to the Patrick Miller Jnr correspondence collected in the National Library for Scotland at MSS 1055. The correspondence by and to Bell in the Seaforth papers in the Scottish Record Office in GD46 is valuable not only for the light it sheds on his Highland interests but for the chance to read, albeit with some difficulty, unedited and unimproved letters by Bell.

For a man who clearly had considerable difficulties with spelling and grammar, Bell was a fairly active writer and it may be of use to provide a chronological list of his appearances in print, so far as they have been traced.

1806 *To the Committee and other subscribers interested in supplying the City of Glagow with water* Glasgow (Broadsheet)

1813 *Observations on the Utility of Applying Steam Engines to Vessels etc* Glasgow (Pamphlet)

1816 Letter on steam navigation in *Caledonian Mercury* 9th October

1817 Letter *Plan of a communication by steam boats between Leith and Greenock* in *Scots Magazine* February

1820 *Letter to the Lord Provost of Glasgow . . . containing the outline plan for the improvement of the Harbour* (Pamphlet)

1825 Letter on steam navigation from *Manchester Guardian* reprinted in *The Glasgow Mechanics Magazine* February

1825 Letter "On the late lamentable accident at Gourock . . ." in *The Glasgow Mechanics Magazine* October

1826 "To the Noblemen, Gentlemen and Freeholders of the County of . . ." (Broadsheet)

1830 "To the Gentlemen, Freeholders, and Merchants of Argyle" Open letter on the Kintyre Canal scheme cited in Edward Morris
Bell may also have been responsible for the following document:

1820 Comet Steam Boat Company: Constitutional Resolutions . . . (Broadsheet)

INDEX

Note: Due to the frequency of their occurence, the entries under the name Henry Bell have been indexed only in connection with his role as Provost of Helensburgh and as the author of various pamphlets, letters etc. Similarly the *Comet* Steamboat of 1812 has not been indexed, except in connection with the Comet Steam Boat Company and the *Comet* Centenary Celebrations of 1912. References to Glasgow are indexed under a particular part of the city, e.g. Broomielaw or under a specific street of building e.g. Clayslap Mill.

A
Admiralty, Board of 113-123
Albion Flour Mills, London 57
Anderson, Alexander (Bank
    agent, Inverness) 177, 201-202
Anderston Secession Church 61
Argyll (2nd.) Paddle-steamer 209
Argyll later *Thames* Paddle-
    steamer 164-169
Ayr Paddle-steamer 204-209

B
Bain, Robert (Captain of *Comet*
    and *Comet II*) 98, 154, 202,
    204, 256
Bain, William (Captain of *City of
    Edinburgh* steam packet) 123,
    134
Barrow, Sir John (Secretary of the

Admiralty) 128
Baths Inn, Helensburgh 21, 60-
    61, 64, 101-112, 139, 147, 227,
    235-236, 238
Bell, Ann (HB's sister-in-law)
    237
Bell, Family 45-47
Bell, Henry, Provost of
    Helensburgh 96-100
Bell, Henry (H.B.'s nephew) 237
Bell, Henry (H.B.'s uncle) 52
Bell, Henry Letter "On the late
    lamentable accident at
    Gourock. . ." in *Glasgow
    Mechanics Magazine* 204, 206-
    207
Bell, Henry Letter on steam-
    boats *Caledonian Mercury* 120-
    122, 126-127, 131

Bell, Henry Letter on steam-navigation in *Glasgow Mechanics' Magazine* 151-152

Bell, Henry *Letter to the Honourable the Lord Provost of Glasgow...* 220-223

Bell, Henry *Observations on the Utility of Applying Steam Engines to Vessels etc* 34-35, 113-116, 120-122, 135-136, 144

Bell, Henry *Plan of Communication by Steam Boats between Leith and Greenock* in *Scots Magazine* 157-159

Bell, Henry *To the committee and other subscribers interested in supplying the City of Glasgow with water* 68-70

Bell, Henry *To the Gentlemen, Freeholders and Merchants of Argyleshire* 230-234

Bell, Henry *To the Noblemen, Gentlemen and Freeholders of the County of...* 122, 249

Bell, Patrick (HB's father) 47

Bell, Thomas (HB's brother) 23, 237

Bell or Easton, Margaret (HB's mother) 47, 70

Bell or Young, Margaret (HB's wife) 24, 61-64, 100-112, 222-228, 236-238

Bellshill 56

Ben Nevis Paddle-steamer 203, 215

Birkinshaw Mill 47

Blantyre Mills 222

Bo'ness 53-55, 134, 136, 153

Boag, Thomas, (Shipbuilder) 54

Boulton & Watt, (Engineers) 57-82, 123, 128, 130

Bridgewater, Duke of (Canal promoter) 84, 123

Brigh Mills 48

Britannia Paddle-steamer 150, 156

Brittanicus (Writer in *Greenock Advertiser*) 39-40

Broomielaw 19, 24, 221

Brunel, Isambard Kingdom (Engineer) 37, 227-228

Buchanan, John (Historian) 27, 35, 64, 140

Buchanan Court, Trongate, Glasgow 60

Burial of Bell 235

C

Caledonia Paddle -steamer 150

Caledonian Canal 199, 202-203, 211-214

Caledonian Canal Commissioners 214-215

Canals and Canal schemes *see* Caledonian, Crinan, Forth and Clyde, Glasgow, Monkland and Tarbert Canals

Canning, George (Prime Minister) 226-227

Carluke Parish Church 65

Carron Ironworks 80-81

Charles-Philippe Experimental steamship 75

Charlotte Dundas 20, 28, 78, 83-84, 125-126, 138, 143, 153

Clayslap Mill, Partick 66

Cleland, James (Glasgow historian) 37, 40, 88-89, 250-251

Clyde Iron Works, Tolcross, Glasgow 222-223

Clyde Navigation Trust 111, 221, 224, 228

Clyde Paddle-steamer 36, 115, 149-150

Coach services 23, 147-148, 156, 159

Colquhoun, Sir James of Luss (Landowner) 95-97

Comet Centenary Celebrations 196, 261-262

Comet II Paddle-steamer 196-200, 202-210, 214
Comet Steam Boat Company 186-189, 192, 196-198
Constable, Archibald (Publisher) 88
Cook, James (Engineer) 40-41, 71, 165
Craignish 192-195
Crinan Canal 170-171, 181, 215, 231-234

D
D'Abbans, Marquis de Jouffroy (Steam-boat experimenter) 75
Dalmonach Mill, Vale of Leven 65, 102, 146, 218
Dalswinton Steam-boat 79-80
Delametherie, J C (Traveller) 168-169
Demologos or *Fulton* US Steam frigate 120-121, 131
Dodd, George (Captain of *Thames* paddle-steamer) 166-168
Douglas, Robert (Captain of *Waverley* paddle-steamer) 235
Duke of Wellington, Paddle-steamer (see also *Highland Chieftain*) 156
Dumbarton 156
Dumbarton Castle Paddle-steamer 150
Dundas, Lord (Governor of Forth & Clyde Canal Company) 83
Dykes, Janet *see* Hamilton, Janet
Dykes, Thomas of Burnhouse 61

E
Easton, Robert (Farmer) 70
Edinburgh 160
Elizabeth Paddle-steamer 36, 115, 148-149, 152, 154
Engines of *Comet* and *Comet II* 195-196

F
Falkirk Burgh School 48
Fingal Paddle-steamer 184
Finlay, Kirkman MP 151
Fitch, John (Steam-boat experimenter) 76
Fly-boats 23-24, 148, 152
Flyter, Robert (Treasurer Comet Steam Boat Company) 186, 188, 214
Fort William 170-172, 174-175, 181-182, 184
Forth, River 152-154, 160-162, 198-199, 239
Forth & Clyde Canal 55, 80-81, 153, 157-160
Forth & Clyde Canal Company 83-84, 153, 160, 229-230
Fulton, Robert (Steam-boat experimenter) 41, 82, 84, 86, 90, 92, 93, 120-121, 123-133, 141, 241-243, 248-250, 253

G
Galt, John *The Steamboat* 104-106
Girdwood, Claud (Engineer) 196
Glasgow Paddle-steamer 37, 41, 149, 156
Glasgow Canal Scheme 220-223
Glasgow Harbour Plans 220-223
Glengarry (*see* Macdonnell of Glengarry)
Golborne, John (Engineer) 23
Gorbals 60
Grangemouth 160
Grangemouth Paddle-steamer 162
Grants & awards to Bell 224-227
Greenock 19, 21, 24, 157-159, 171, 203, 208
Greenock Paddle-steamer 150-151, 156, 169

H
Hamilton or Young or Dykes, Janet 61, 107
Hardie, Thomas (Engineer) 30

Harriston, William *The Steam-Boat Traveller's Remembrancer* 38
Hart, Alexander (Shipbuilder) 83
Hebridean Steam-boat project 172-183
Helensburgh 21-22, 30, 95-112, 147-148, 156, 171, 237-238
Helensburgh Town Council 96-100
Helensburgh United Presbyterian Church 237-238
Henry Bell Paddle-steamer 224
Highland Chieftain (see also *Duke of Wellington*) Paddle-steamer 185, 210
Highlander Paddle-steamer 203
House of Commons, Select Committee on the Holyhead Roads 166, 172, 248-249
Hulls, Jonathan (Steam-boat experimenter) 74-75

I
Incorporation of Wrights 59-60
Inglis, James (Engineer, Bellshill) 56
Ingram Street, Glasgow 60
Innes, Sir Hugh (Landowner) 172-177, 181, 183, 200
Inveraray 169

J
Jay Mill 52

K
Kelvin, River 222
Kibble, James (Mill-owner) 146
King Street, Glasgow 60
Kinneil Mill, Bo'ness 54

L
Leith 153, 157-159, 239
Leven, River 217-219
Lewis 173-176
Livingston, Robert (US Ambassador to France & steam-boat entrepreneur) 124, 131-132
Loch Lomond 217-219
Logan, Walter (Superintendent of Forth & Clyde Canal Company) 229-230

M
McClelland, Thomas (Captain of *Ayr* paddle steamer) 208
Macdonnell of Glengarry, Col Alexander, (Landowner) 211-213
McGeorge, Andrew *Old Glasgow, the place and the people* 24
McInnes, Duncan (Captain of *Comet II* paddle-steamer) 208
McInnes, Duncan (Pilot of *Comet*) 35
MacKenzie, Captain William (First master of *Comet*) 27-28, 35, 154
Mackenzie, J A Stewart of Seaforth (Landowner) 172-173, 176-183, 201, 203
Mackenzie, Peter *Old Reminiscences of Glasgow...* 23, 37
MacLeod, Donald *A nonogenerian's reminiscences...* 99, 145, 219-220, 234-235, 254-255
Margery later *Elise* Paddle-steamer 165-166
Martin, Alexander & Coy (Shipbuilders, Port Glasgow) 164-165
Maugham, William *Annals of Garelochside...* 138, 218-219
Meikle, Andrew (Millwright) 52
Melville, 1st Viscount (Henry Dundas) (1st Lord of the Admiralty) 113-119, 122, 124
Melville, 2nd Viscount (Robert Saunders) (1st Lord of the

Admiralty) 88, 118, 129
Melville Court, Glasgow 60
Melville Place, Glasgow 60
Merchants House of Glasgow
    225-226
Miller, Patrick (Landowner and
    steam-boat experimenter) (*see*
    also Dalswinton Steamboat)
    20, 76-82, 129-130, 137, 143-
    144, 243-245
Miller, Patrick Jnr (Landowner)
    50, 86-90, 129-130, 241-243
Milligs (*see* also Helensburgh) 95
Mitchell, Joseph (Transport
    engineer) 199-200, 202-203
Monklands Canal 223
Montrose, Duke of (Landowner)
    218-219
Monuments to Bell 235-236, 247-
    248
Morris, Edward *The Life of Henry
    Bell* 15, 39-40, 42-43, 45-46, 50,
    60, 112, 116, 127, 144-145, 146,
    170-171, 200, 220, 224, 226-
    228, 234
Munro, Neil *The New Road* 213

N
Napier, David (Engineer and
    builder of boiler for *Comet*)
    27, 29-30, 66, 70, 191, 247
Napier, Robert & Sons
    (Shipbuilders and engineers)
    196
Napier, Robert (Engineer) 247-
    248
Nautilus Submarine 121, 123-124,
    133
Nelson, Admiral Horatio 1st
    Visc 116-119, 122
Newbigging, Archibald
    (Merchant) 63-64, 102
Newhaven 160, 162
Nicol, James (Shipwright) 30,
    171
North River Steam-boat 131, 141

P
Pagan, J *Glasgow Past and Present*
    24
Paine, Thomas (Writer) 75-76
Paisley 151
Papin, Denis (Steamboat
    experimenter) 75
Paterson, James (Business
    partner of HB) 59-60, 65
Port Glasgow 27, 30
Prince of Orange Paddle-steamer
    150-151
Princess Charlotte Paddle-steamer
    150
Pyroscaphe Experimental
    steamship 75

Q
Queen's Hotel (*see* Baths Inn)

R
Raeburn, Sir Henry (Artist) 212
Railways 223
Rennie, Andrew (Inventor of
    Rennie's Wheel-Boat) 138
Rennie, John (Engineer) 43, 52,
    56-58, 82, 128
Robertson, John (Engine-builder
    of *Comet*) 27-28, 31-32,34-35,
    42, 64, 140, 143, 147, 190, 195-
    196, 252
Robertson, Robert (*Comet*'s
    Engineer) 35

S
Sallachan Point 191-192
Science Museum 195-196
Scott, Sir Walter (Novelist) 212
Scottish Maritime Museum 196
Scotts Shipbuilding & Engineer-
    ing Coy , Greenock 60
Secession Church 61, 237-238
Shaw & Hart (Shipbuilders) 54-
    55, 134, 136, 153
Shipwrecks 192-196, 204-212
Smith, James of Jordanhill

(Provost of Helensburgh) 102, 111

Stanhope, Earl (Steamboat experimenter) 57, 82, 123, 128, 132

Steamships, historical development 74-76

Stirling or *Stirling Castle* Paddle-steamer 177, 198-201, 203-204, 210-213, 249

Stonebyres Linn, Lanarkshire 68-69

Stowe, Harriet Beecher (Novelist) 110

Symington, William (Engineer) (see also *Charlotte Dundas* and Dalswinton Steamboat) 78-81, 83-88, 90, 92, 125, 129, 241-243, 245-246

T

Tarbert Canal 230-234

Taylor, James (Tutor to Patrick Miller & steamboat experimenter) (*see* also Dalswinton Steamboat) 76, 78-82, 84, 86-87, 90-92, 125, 137-138, 143-144, 240, 243-246, 255

Telford, Thomas (Engineer) 24, 67-69, 200

Thames ex *Argyll* Paddle-steamer 164-169

Thomson, John (Engineer & shipowner) (*see* also *Elizabeth*) 36, 135-145, 152-154, 184

Thomson, William (Resident engineer at Crinan Canal) 170-171, 173, 196-198, 207

Torphichen 47-48

Tourism 37-38, 104-110, 156, 164-169, 203-204

Trusty Paddle-steamer 149

Tug Paddle-steamer 160-161

W

Water-supply schemes 67-69, 98-100, 217-219

Watt, James (Engineer) (*see* also Boulton & Watt) 20, 57, 82, 137, 252-253

Waverley Paddle-steamer 235

Wood, John (Ship-buillder, Port Glasgow) 27-30, 34, 37, 146, 161

Wright, Dr Archibald (Shareholder in Comet Steamboat Company) 51, 188-190

Y

Young, Janet (*see* Hamilton, Janet)

Young, Margaret (*see* Bell, Margaret)

Young, Robert (Father of Margaret Young or Bell) 61